T £4.50

Digital Hardware Design

Digital Hardware Design

Ivor Catt
David Walton
Malcolm Davidson

M

First published 1979 by
THE MACMILLAN PRESS LTD
London and Basingstoke
Associated companies in Delhi Dublin
Hong Kong Johannesburg Lagos Melbourne
New York Singapore and Tokyo

Typeset in 10/12 Times by
Reproduction Drawings Ltd Sutton, Surrey
and printed in Great Britain by
Unwin Brothers Limited
The Gresham Press
Old Woking, Surrey

British Library Cataloguing in Publication Data

Catt, Ivor
    Digital hardware design.
    1. Digital electronics
    I. Title   II. Walton, David, b.1947
    III. Davidson, Malcolm
    621.3815   TK7868.D5

    ISBN 0-333-25981-5

Contents

Preface

The authors, who have long experience of research and development in digital electronics, became concerned at the lack of attention paid to the design of digital hardware by educational establishments. This book is based on material developed for a series of seminars which are now attended by engineers drawn from the design laboratories of many large electronics companies in the United Kingdom.

If and when colleges and faculties of electronics and physics realise that digital hardware design is a major discipline, this and subsequent volumes planned by the authors will be invaluable for the design of relevant and up-to-date courses.

I. CATT
D. S. WALTON
M. DAVIDSON

1

Introduction

All engineering is based on science, as academic establishments quite rightly emphasise. However, it is essential that the science which is taught is relevant to current practice and technology in engineering. If this is not so, the student comes to industry armed with a set of scientific principles, and corresponding mathematical techniques, which are inappropriate to his work as a designer and which lead to disillusionment with science in general. Worse still, the diligent application of inappropriate techniques may lead to bad design, often with disastrous consequences.

The digital electronics industry has sprung up so quickly in the last 10 years that the theoretical foundation required has not developed at all. It is impossible to cross the line separating the analog and digital worlds. The sine wave is a periodic time-varying steady state situation, whereas a digital signal is a fixed amplitude step (shock wave). Each change of state is a single event in time and cannot be correlated with any other change. A dubious connection, via Fourier analysis, is merely a mathematical *arpeggio*, guaranteed to be worth a few exam questions at least. A leading edge of a step is a shock wave; it is a transverse electromagnetic wavefront which travels at the speed of light. Of course, it is possible to take this single step and analyse it using Fourier analysis but this would mean combining an infinite number of sine waves which exist from minus infinity to plus infinity. This can be easily seen to be quite absurd and of no practical use.

At the turn of the century Oliver Heaviside and his contemporaries Lodge, S. P. Thompson and Hertz developed many theories which

1

should be used today. By thinking of digital signals as small discrete packets of 'energy current' flowing at the speed of light between the wires (which merely act as a guide) many of the present-day design implementation problems could be solved. The advent of the telephone and radio led to the predominance of sinusoidal time-varying signals, so the concept of 'energy current' was lost as new theories were developed to cope only with the periodic waveform. We have now turned a full circle and must look backwards before we can advance. The works of Oliver Heaviside contain a huge amount of relevant information and any serious minded digital engineer would gain enormously by reading his books.

The practical problems related to digital systems such as

(1)   crosstalk (noise)
(2)   power supply decoupling
(3)   signal termination and drive techniques
(4)   component pulse response
(5)   grounding
(6)   line-borne interference

need to be studied. General models and original concepts based on Heaviside's 'energy current' idea can be used to tackle the above problem areas making it possible to design complex digital systems in an orderly scientific fashion. Every practising engineer in digital electronics must stop attempting to use analog ideas for digital systems; they will not work. This is easy to see; all around the industry are scattered systems which 'crash' regularly. Pattern sensitivity, noise, power supply problems are all raising their ugly heads, and all quite unnecessarily. By following clearly defined design rules, systems can be built which will work reliably and first time, without the usual 3 to 6 month commissioning troubles. It is difficult to assess the financial saving that could be made if digital systems were developed using adequate theoretical principles. Suffice it to say that the saving would be significant. Also the job satisfaction of development teams would increase.

The hard and fast rules laid down for periodic sine wave situations must be cast aside and new rules developed for the shock wave situation. An obvious area to concentrate on is the one of signal distribution. Any prime source of electrical energy, be it analog or digital, needs to be easily distributed to loads that require it. We must have a basic understanding of the mechanism by which a block or pulse of energy is transmitted in space. This leads us into the realms of electromagnetic field theory, for it is here that the student will learn and ultimately understand the subject of digital electronics.

Unfortunately, nearly all the books written on the subject of electromagnetic field theory are concerned with steady state sine wave situations. There is no basic theory written today which concentrates on high speed digital techniques. The knowledge of how 1 ns steps propagate is known by only a few people. Yet with the advent of ECL (emitter coupled logic) and Schottky TTL this electrical phenomenon is becoming widespread. Engineers today attempt to put together fast, complex logic systems which are doomed to failure. The paper design might well be satisfactory but the problems that arise during testing and commissioning seem endless. The unfortunate engineer just cannot understand the 'gremlins' that keep upsetting his system. This is because nowhere is he taught the important fundamental principles necessary for competent digital system development.

In order to have a complete understanding of high speed systems one must apply certain techniques which are not taught in any educational establishment, nor written about in any textbook. One must go back to the turn of the century to find any suitable material. Then, the main subject area was telegraph signalling which is analogous to digital transmission today. A 10 ms risetime step or edge travelling 1000 km (telegraphy) is based on the same theoretical principles as a 1 ns step travelling 10 cm (computers).

Finally, and probably the most important point, not one of the design concepts that are used is difficult. Although soundly based in theory, they do not involve exotic mathematics and are aimed specifically at practical problems of hardware development. They are tools of the trade to be used by all engineers and technicians. There is no need to allow ourselves to be surrounded by a fog of complex but inappropriate mathematics, when there is the chance to gain a clear understanding of a challenging, high technology industry.

2

The Interconnection of Logic

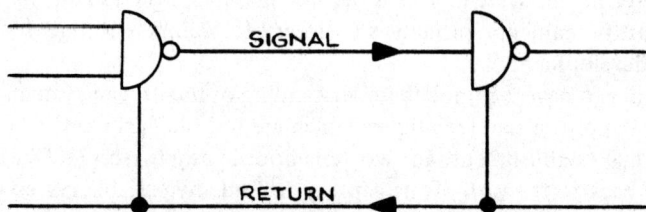

Figure 2.1

A logic system is composed of logic gates and interconnections between them. It is important, in order to ensure the correct functioning of the system, that an adequate model of this connection be used.

The simplest case is that of a logic gate driving another single gate, that is, there is no fanout. We shall return to the case of fanout in a later section.

It is impossible to consider the interconnection in the absence of a ground (or V_{cc}) return path. When this is present we have both distributed capacitance and distributed inductance or, in other words, a *transmission line*. Thus the appropriate model of interconnection between two gates is a transmission line.

Properties of a Transmission Line

In order to discover how we characterise a transmission line we shall consider a step propagating along a two-wire line (see figure 2.2).

4

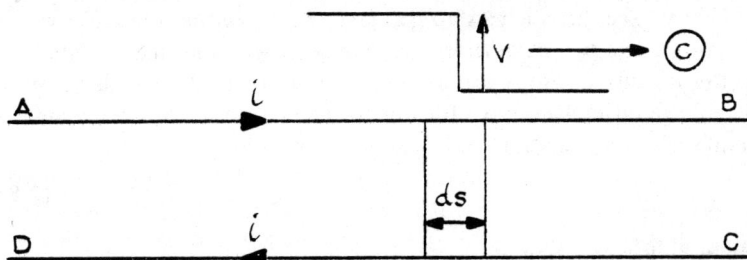

Figure 2.2

Use Faraday's law ($V = -d\phi/dt$) around the loop ABCD.

Define L as the inductance per unit length of the wire pair, then

$$L = \frac{\phi}{i} \tag{2.1}$$

In a time t, the step will advance a distance δs, such that

$$\frac{\delta s}{\delta t} = c \tag{2.2}$$

and the change of flux will be (from equation 2.1)

$$\delta\phi = L \, \delta s \, i \tag{2.3}$$

Substitution into Faraday's law gives the voltage applied to the line to overcome the back e.m.f.

$$v_{AD} = Li \frac{ds}{dt} = Lic \tag{2.4}$$

Now we also obtain, from the definition of a capacitor $q = vC$

$$i = vCc \tag{2.5}$$

where C is the capacitance per unit length of the wire pair.
  It now follows that

$$c = \pm \frac{1}{\sqrt{(LC)}} \tag{2.6}$$

and also

$$\frac{v}{i} = Z_0 = \sqrt{\left(\frac{L}{C}\right)} \tag{2.7}$$

Thus we see that such a step may propagate in either the x or $-x$ directions. The two parameters that characterise a transmission line are the velocity of propagation c and the impedance Z_0, which relates the voltage difference across the line to the current in the line by an 'Ohm's law' type relation

$$v = iZ_0 \tag{2.8}$$

where Z_0 is a property of

(1)   the geometry of a cross-section of the wires and
(2) μ and ϵ for the medium in which the wires are embedded; this will be shown in chapter 3.

It should be noted that the above argument is independent of the idea of frequency, and Z_0 and c do not depend on frequency provided μ and ϵ are frequency-independent, as they are in practice.

In order to use the formulae for Z_0 and c we must have methods for determining L and C for any geometry we might be using. In general it will be impossible to solve analytically for L and C and so other methods must be resorted to.

3

The Analogy between L, C and R

In this chapter we develop a useful analogy which leads to simplified calculation in all cases and to a simple technique of measurement in those cases which do not yield to calculation.

We shall consider the special case of a parallel-plate transmission line (see figure 3.1).

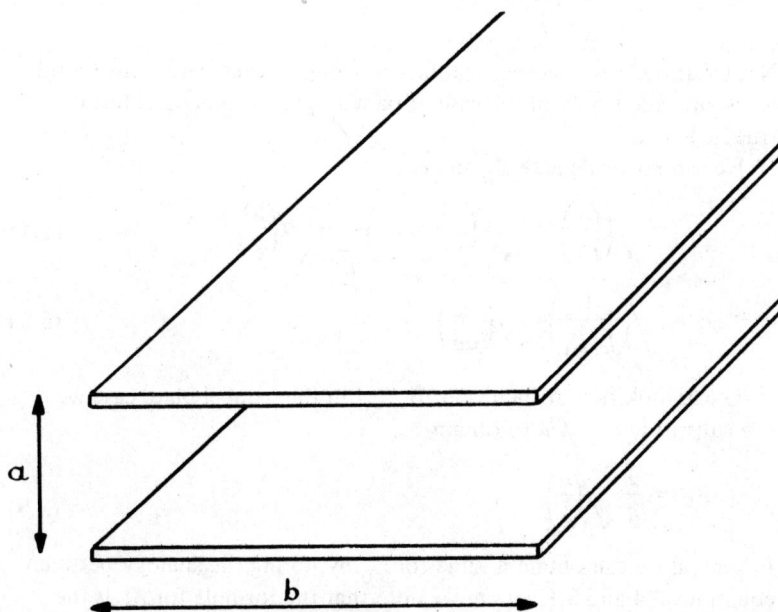

Figure 3.1

We shall calculate the resistance, capacitance and inductance per unit length of the line

Resistance If the medium between the plates has resistivity ρ, then the resistance between the plates per unit length is

$$R_1 \;=\; \rho \frac{a}{b} \;=\; \rho \frac{1}{f} \tag{3.1}$$

where we have defined $b/a = f$ as a geometrical factor which is a dimensionless function of the dimensions of the line.

Capacitance For a parallel-plate capacitor

$$C_1 \;=\; \epsilon \frac{b}{a} \;=\; \epsilon f \tag{3.2}$$

Inductance

$$L_1 \;=\; \mu \frac{a}{b} \;=\; \mu \frac{1}{f} \tag{3.3}$$

Note that the same geometrical factor occurs in each case. This useful result holds not only in the case of parallel-plate geometries, but is true in general.

We can now calculate Z_0 and c.

$$Z_0 \;=\; \sqrt{\left(\frac{L_1}{C_1}\right)} \;=\; \sqrt{\left(\mu \frac{1}{f} \times \frac{1}{\epsilon f}\right)} \;=\; \frac{1}{f}\sqrt{\left(\frac{\mu}{\epsilon}\right)} \tag{3.4}$$

$$c \;=\; \sqrt{\left(\frac{1}{L_1 C_1}\right)} \;=\; \sqrt{\left(\frac{1}{\mu\epsilon}\right)} \tag{3.5}$$

Let us look first at the result for Z_0. In the parallel-plate case we can substitute $f = b/a$ to obtain

$$Z_0 \;=\; \frac{a}{b}\sqrt{\left(\frac{\mu}{\epsilon}\right)} \tag{3.6}$$

In general we can obtain a value for Z_0 by noting the analogy between equations 3.4 and 3.1, where we note that the formula for Z_0 is the same as that for R_1 except that ρ has been replaced by $\sqrt{(\mu/\epsilon)}$. This

means that we can obtain the geometrical factor by calculating the
resistance between the conductors and multiplying by the factor
$(1/\rho) \sqrt{(\mu/\epsilon)}$. In cases where a calculation cannot be made, measurements
using resistive paper can be used. Here the conductors are painted on
to the resistive paper using conducting paint and the resistance between
them measured using an ohmmeter. The equivalent to the resistivity
is the resistance between two sides of a square of this paper.

Note also the result of equation 3.5, which shows that the
velocity of a wavefront down a two-conductor system is independent
of the geometry (provided this is two-dimensional) and is a property
only of the medium in which the conductors are placed.

We shall now use the results just derived to obtain the impedance
of a co-axial line.

Impedance of a Co-axial Line

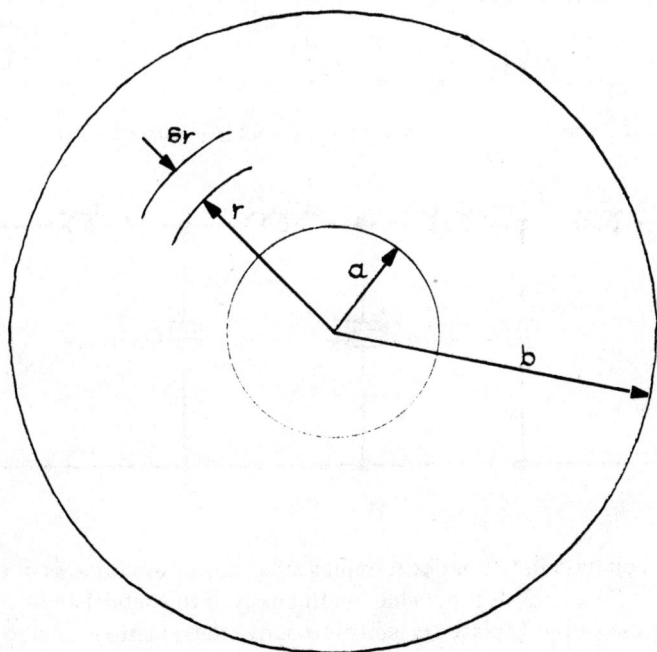

Figure 3.2

Consider figure 3.2. Outwards resistance of a co-axial shell at radius r is

$$R_r = \rho \frac{\delta r}{2\pi r} \tag{3.7}$$

Therefore the resistance between inner and outer conductors for a unit length of cable is

$$R_1 = \int_a^b R_r \, dr = \frac{\rho}{2\pi} \int_a^b \frac{dr}{r} = \frac{\rho}{2\pi} \ln \frac{b}{a} \tag{3.8}$$

Thus in this case, the geometrical factor is

$$f = \frac{2\pi}{\ln \frac{b}{a}} \tag{3.9}$$

and therefore the impedance is

$$Z_0 = \frac{1}{f} \sqrt{\left(\frac{\mu}{\epsilon}\right)} = \frac{1}{2\pi} \ln \frac{b}{a} \sqrt{\left(\frac{\mu}{\epsilon}\right)} \tag{3.10}$$

which is the standard result for the impedance of a co-axial line.

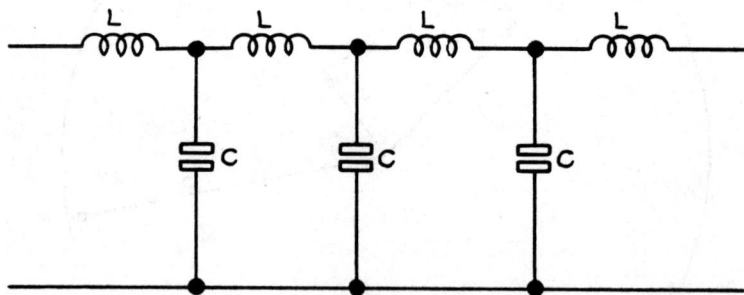

Figure 3.3

It is common in textbooks to represent a transmission line as shown in figure 3.3. It is possible, on the basis of this model and making use of the Laplace transform to derive the equations of step propagation. However, this method has little to recommend it especially since it appears to lead to a high frequency cutoff which is quite spurious. There is of course no high frequency cutoff inherent in any transmission line geometry. The only factor which can lead to high frequency cutoff is a frequency-dependent behaviour of the dielectric. Clearly, if the dielectric is a vacuum there is no high frequency cutoff.

Effect of Discontinuities in Transmission Lines

Consider a junction between transmission lines of different impedance in figure 3.4.

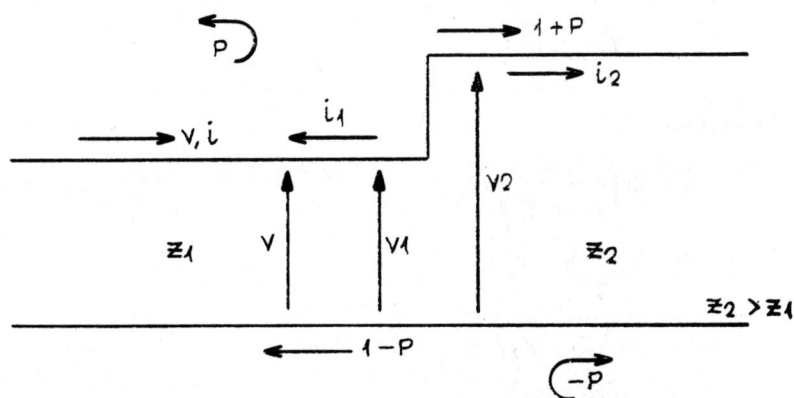

Figure 3.4

Consider a step v, i incident from the left along line 1. Suppose that this results in two waves at the junction: v_1, i_1 travelling back to the left and v_2, i_2 travelling onwards towards the right. We have

$$v = iZ_1 \tag{3.11}$$

Conservation of current (Kirchhoff's first law) gives

$$i = i_1 + i_2 \tag{3.12}$$

For voltage equality across the junction of the two lines we must have

$$v + v_1 = v_2 \tag{3.13}$$

Now we have the relationships

$$v_1 = i_1 Z_1 ; \quad v_2 = i_2 Z_2 ; \quad v = i Z_1 \tag{3.14}$$

and equations 3.11, 3.12 and 3.13 give

$$\frac{v}{Z_1} = \frac{v_1}{Z_1} + \frac{v_2}{Z_2} \qquad (3.15)$$

Substituting $-v_1 = v - v_2$ from equation 3.13 into equation 3.15 gives

$$\frac{2v - v_2}{Z_1} = \frac{v_2}{Z_2}$$

$$\frac{2v}{Z_1} = v_2\left(\frac{1}{Z_1} + \frac{1}{Z_2}\right)$$

$$2v = v_2\left(\frac{Z_1 + Z_2}{Z_2}\right)$$

Therefore

$$v_2 = 2v\left(\frac{Z_2}{Z_1 + Z_2}\right) \qquad (3.16)$$

Now since $v_1 = v_2 - v$

$$v_1 = 2v\left(\frac{Z_2}{Z_1 + Z_2}\right) - v$$

$$v_1 = v\left(\frac{Z_2 - Z_1}{Z_2 + Z_1}\right) \qquad (3.17)$$

If we define a reflection coefficient ρ such that $\rho = v_1/v$, then

$$\rho = \frac{Z_2 - Z_1}{Z_2 + Z_1} \qquad (3.18)$$

and we can also write

$$v_2 = v(1 + \rho) \qquad (3.19)$$

For an incident wave to the right, the relevant coefficients are $-\rho$ for the reflected wave and $(1 - \rho)$ for the wave continuing to the right.

Open and Short Circuit Lines

If we have a transmission line which comes to a dead end which is an open circuit, we can think of the open circuit as being a line of infinite impedance.

Equation 3.18 shows that as Z_2 approaches infinity, ρ approaches 1. Thus the step is reflected from the open circuit without inversion. This means that immediately the step arrives at the open circuit it is reflected and adds itself to the voltage already on the line, leading to the familiar doubling effect. It is possible to observe the effect by using an oscilloscope with an input impedance much greater than the line impedance.

Consider now a line which ends in a short circuit, that is, $Z_2 = 0$. In this case $\rho = -1$, and the step is reflected with inversion so that no voltage is seen at the shorted end. This is hardly surprising since it is a short circuit.

Termination of Lines

We see that in both cases considered above, the step was reflected back into the line without loss of energy. In practice, when we interconnect logic this is an undesirable effect. We would like the step to be perfectly absorbed at the receiving end of the line so that it will have no further consequences.

We see how to achieve this if we note that the mathematics relating to the discontinuity in the line can still be carried through if Z_2 is replaced by a resistor. If we now adjust the value of the resistor to be equal to Z_1, we see that $\rho = 0$ and there is no reflection. In a sense there is now no discontinuity in the line and the step behaves as though it were traversing an infinitely long line, that is, it disappears!

Power Loss at a Discontinuity

We shall consider a step traversing from a line of impedance Z_0 to one of impedance $2Z_0$.

In this case, $\rho = 1/3$.

If the incident step is of voltage v, its power is v^2/Z_0.

Beyond the change of impedance, the voltage of the step increases to

$$v(1 + \rho) = \frac{4v}{3}$$

and the power becomes

$$\frac{16v^2}{9} \times \frac{1}{2Z_0}$$

that is, $^8/_9$ of the power continues on to the right and only $^1/_9$, or 11 per cent, is reflected back. We notice that after the discontinuity the voltage is increased and the current decreased. The discontinuity therefore behaves as a rather lossy transformer.

In practice, Z_0 is only a slowly varying function of the geometry of the line, and hence even quite large variations in the geometry lead to insignificant reflections (see figure 3.5).

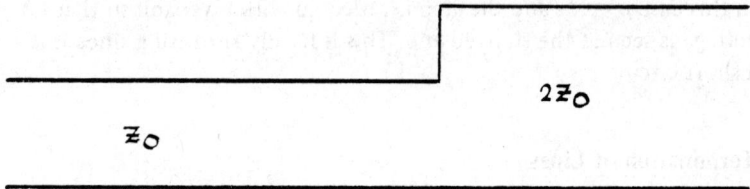

$2Z_0$

Z_0

Figure 3.5

4

Transmission Line Theory Applied to Logic Interconnection

We have seen in the previous chapter that the interconnection between logic gates should be viewed as a transmission line, and have examined the properties of such lines. We shall now look at the sort of device which would be ideal for driving and receiving pulses on such a line, and then consider how closely the available logic families approach this ideal.

It is assumed in this chapter that we are dealing with high speed logic, and that the reduction of unnecessary delays is our prime requirement.

Transmission Line Driving

We have already noted that, in order to avoid reflections, a transmission line should be terminated in its characteristic impedance. We shall examine the implications of this consideration in the case of TTL (transistor–transistor logic). The operation of the basic TTL gate is described elsewhere. Here we shall only look at its input and output circuits (see figure 4.1).

Historical Summary

Before considering the TTL circuit in detail we shall spend some time considering the evolutionary process which led up to it.

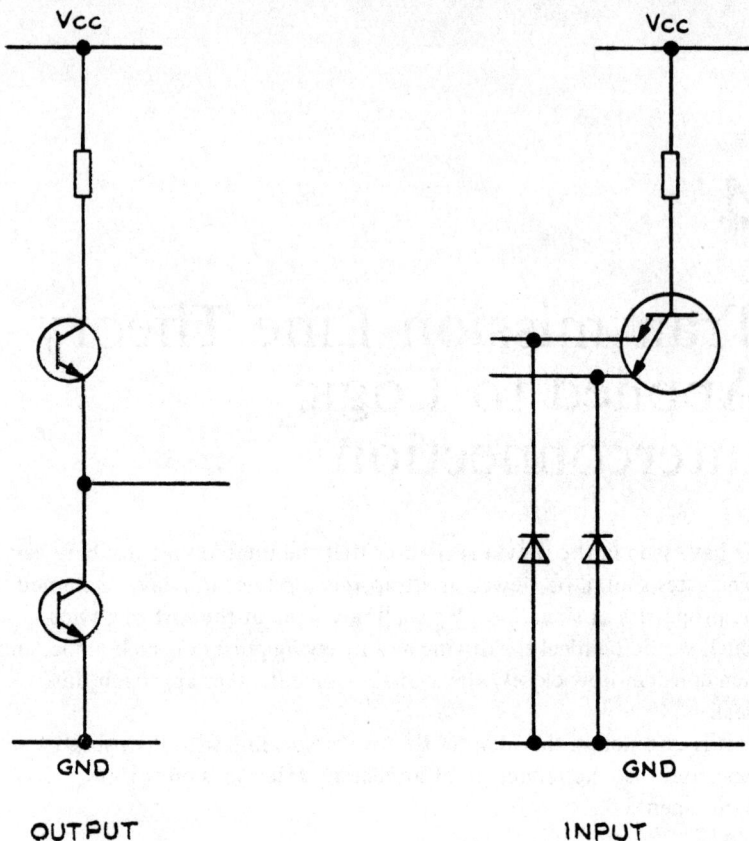

Figure 4.1

In early DTL (diode–transistor logic), transistors were only capable of sustaining 1 mA collector currents. This led to the circuit arrangement shown in figure 4.2 where a 10 kΩ resistor was used. The circuit shown is clearly a NOR gate, which was the basic DTL element.

One of the problems noted in practice with this circuit was its inability to drive stray capacity. Consider the output waveform in figure 4.3, which is obtained when a pulse is applied to the input.

When the transistor switches on, the stray capacitance C is discharged through the saturated transistor which offers an impedance

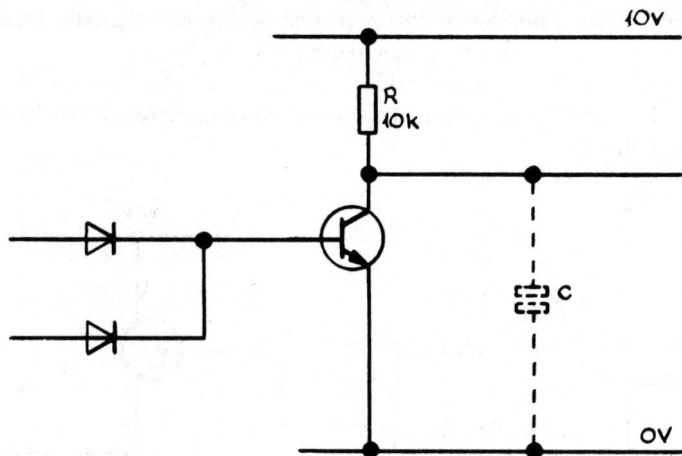

Figure 4.2

much less than 10 kΩ; the falling edge is therefore very rapid. The rising edge, however, presents a different picture. Here the transistor switches off and is unable to supply current to charge the stray capacitance through the reverse-biased base-collector junction. This means that the stray capacitance must be charged through the 10 kΩ resistor, leading to an exponential rise which corresponds to a time constant of RC. This is clearly unacceptable since it means that the gate is very poor at driving long signal lines.

Let us, however, recall that in practice the load is not strictly capacitive, but is in fact a transmission line of characteristic impedance around 100 Ω. It is then instructive to ask whether this makes any difference to the conclusion. The answer is that it makes

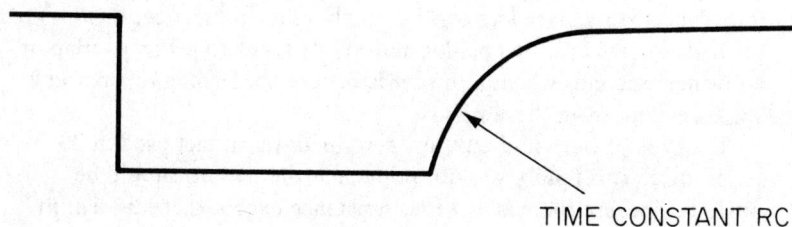

TIME CONSTANT RC

Figure 4.3

no appreciable difference in this case because R is much greater than
100 Ω; in this limit the capacitive model is quite good.

Designers of logic families tried to circumvent this problem with
a two-way 'push–pull' output stage which would give rapid transitions
in both directions.

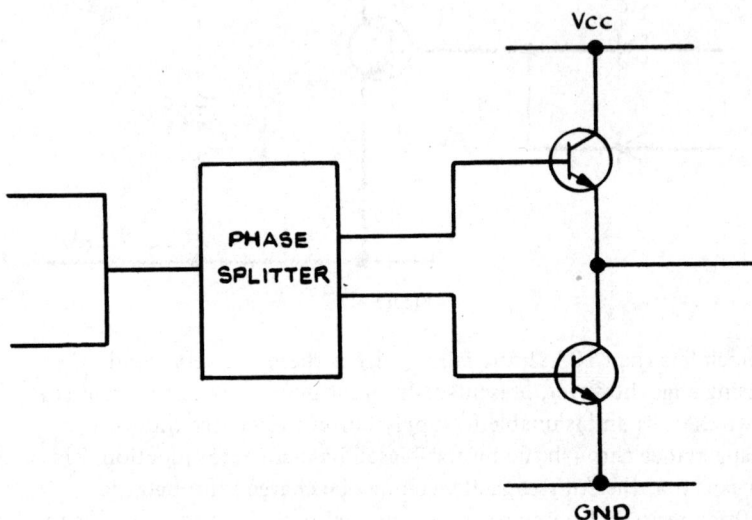

Figure 4.4

They ended up with a totem pole circuit driven by a 'phase
splitter' (terrible misnomer) which would ensure that the top transistor
was on for a high and the bottom one off, and vice versa (see figure 4.4).
One should clearly avoid the disastrous situation of having both
transistors turned on simultaneously, which leads to a very low
impedance being placed across the supply rails. In practice, it is
difficult to avoid this happening and so one tends to get an overlap of
a few nanoseconds when both transistors are on, leading to a current
spike demand from the supply.

The type of output circuit just described was in fact used in 73
series logic. This family was not popular – the reasons should be
obvious. Because there is no series resistance except that inherent in
the semiconductors themselves, the current transients are considerable.

So the final step in the evolution of TTL was the insertion of a
series limiting resistor in the collector of the upper transistor,

leading to the output configuration familiar to us all.

Unfortunately, the history of the subject has gone away from a 'straight and narrow' path because, in the evolution of TTL, insufficient account has been taken of the way impedance levels and device speeds have changed. Thus, while DTL was working initially with a 10 kΩ output impedance, we now have TTL devices where the internal pullup resistor is of the order of 100 Ω. The implication of this is that, while with the higher resistance it was permissible to consider the output as being capacitively loaded, with 100 Ω resistance values this is no longer a sensible model. In fact, since the output load is a transmission line with a characteristic impedance of around 100 Ω, we can no longer neglect its behaviour by lumping the connection and calling it a capacitor.

In fact, it turns out that when we model the interconnection as a transmission line, the upper transistor in the TTL totem pole is redundant and serves no useful function! This is a startling conclusion and takes a little digesting, but if we work through an example, the implications should become clear.

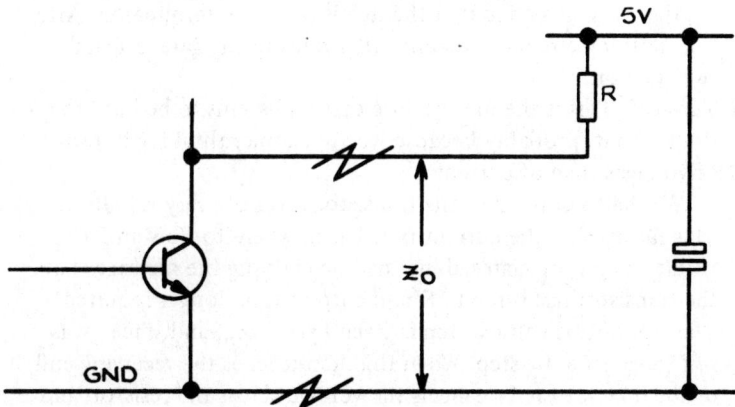

Figure 4.5

Open-collector Gate Driving a Transmission Line

The transistor in figure 4.5 could be (and normally would be) the output transistor of a TTL open-collector gate. It is usually thought that the main function of these gates is to provide a wire AND facility in application where speed is not particularly important; but more of this later.

The transistor is connected to a transmission line of impedance Z_0 and has a pullup resistor R to +5 v at the far end. From the point of view of signals, this R shunts the line and therefore, as we have seen earlier, any step arriving at R will not be reflected.

We shall now consider what happens when the transistor switches on and off.

(1) Turn-on. When the transistor is off there is no current through R, and therefore the signal conductor is at +5 v. When the transistor turns on, its collector voltage falls rapidly to less than 1 v. The time taken for this depends on the logic family, but it could be as short as 1 ns. We shall assume for a moment that whatever this time is, it is much less than the time taken for a step to traverse the line.

After the transistor has switched, a current/voltage step travels along the line. If the voltage step is 4 v, the transistor will sink a current given by $i = 4/Z_0$. When the step reaches the receiving end it will be exactly the right size to draw the same current i through the resistor R if we make $R = Z_0$. This is another way of saying that the line is perfectly terminated. After this, nothing else happens until the transmitting gate is again switched on.

(2) Turn-off. This is the more subtle case and seems to be hard to think about, probably because we are happier thinking in terms of voltages than of currents.

We shall assume that the transistor turns off very rapidly. This means that the current now has nowhere to flow and a voltage step propagates along the line carrying the message that the transistor has turned off and current is no longer required! The size of this voltage step is given by $v = iZ_0$, and since i was $4/Z_0$, we get a 4 v step. When this step reaches the receiving end of the line, it exactly cancels the voltage across the resistor, and the current flow stops.

Thus we see that no active pullup is necessary; open-collector gates will quite happily act as line drivers.

Someone is sure to object that they have observed the waveforms at the output of open-collector gates giving the 'normal' exponential rise type of behaviour, and they are quite correct! The point is that if R is far greater than Z_0, the behaviour of the system approximates more and more to what one would expect if the interconnection were replaced by a lumped capacitor. This is simply because, since the line

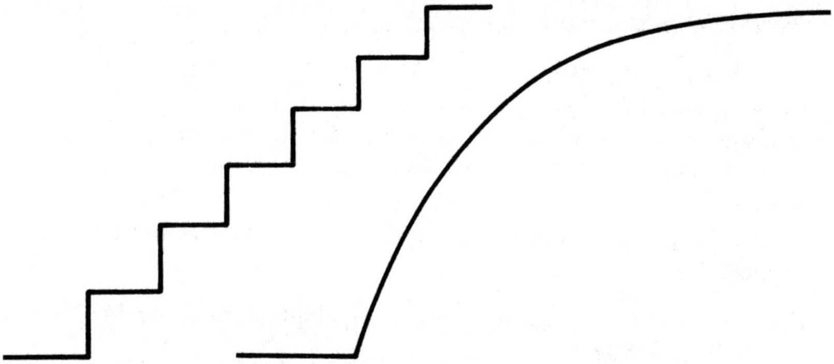

Figure 4.6

is unmatched, the step must make several traverses of the line before it reaches 5 v. In other words, the receiving end behaviour is the staircase in figure 4.6. If this staircase is seen on an oscilloscope of restricted bandwidth, it will be smoothed to give the appearance of the second waveform – a 'capacitive' behaviour.

Thus, in order to obtain fast settling of the line, we must have R approximately equal to Z_0.

It is worth noting that if R is far smaller than Z_0 we obtain something which approximates to 'inductive' behaviour, with an output waveform shown in figure 4.7.

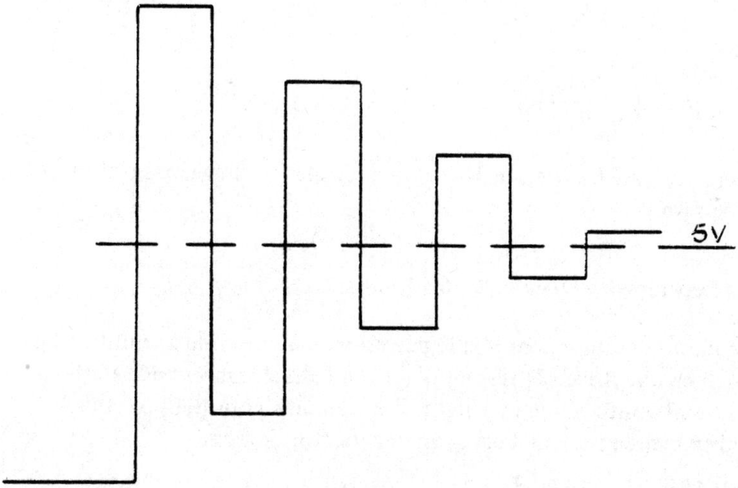

Figure 4.7

Practical Considerations

There is a practical problem when we come to implement line drivers using open-collector gates. This is that an average open-collector device is unable to sink the current it is required to when its output is pulled up to +5 v by a resistor of around 150 Ω (the sort of Z_0 one encounters in twisted pairs, for instance).

There are two solutions to this and in some cases both must be applied.

(1) Use devices with high sink current capability such as the 7417, 7438 or 74S38.
(2) Do not pull up to +5 v but only to about +3 v. This has the effect of slightly degrading the high level noise immunity, but is usually acceptable. Slightly over-terminate.

If the second course is adopted, it is not necessary to provide a separate power supply. Simply terminate the line with two resistors, R_1 to +5 v and R_2 to ground. These resistors are then chosen so that their Thévenin equivalent network has a voltage generator of +3 v and an impedance equal to Z_0

By considering the open circuit voltage and short circuit current it is easily seen that

$$V_T = 5\left(\frac{R_2}{R_1 + R_2}\right)$$

and

$$R_T = \left(\frac{R_1 R_2}{R_1 + R_2}\right)$$

where V_T and R_T are the Thévenin equivalent voltage and resistance respectively.

Bus Driving with Open-collector Gates

A common requirement in a logic system is to provide a common bus which can be driven at several points and which can provide outputs at several points along its length. We shall now consider how this facility can be provided using open-collector devices.

The open-collector device was, until recently, the preferred type
for common bus systems but it has recently fallen into disfavour to be
replaced by the tri-state type of chip. This is a retrograde step.

The open-collector device fell into disfavour unnecessarily. As we
have seen, it is quite capable of driving transmission lines, and does
not need the 'crutch' of the upper transistor to obtain rapid switching
times.

We shall consider a single conductor which forms part of a bus
and see how we would go about driving signals along it (see figure 4.8).

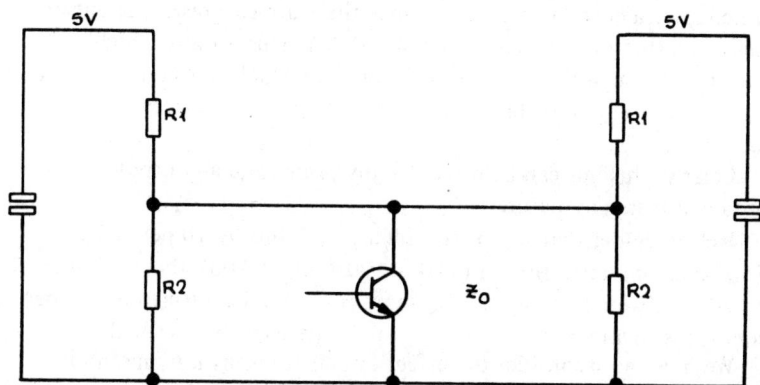

Figure 4.8

For the sake of convenient working, we shall assume that
Z_0 = 100 Ω. The values may be scaled to suit any other value of Z_0.

The driving gate, of which only one is shown, is in fact loaded by
two transmission lines going off in each direction. Provided there is
no interaction between the lines (which there is not in this case since
we assume they go out in opposite directions), the driving gate will
'see' an impedance of Z_0 in parallel with another Z_0 making 50 Ω.

We shall have to terminate the bus at each end so that there will be
no further reflection of any signal arriving there. However, since it is
difficult to sink the current which such a low termination will inflict
on the driving transistor, we shall terminate such that R_T is about
10 per cent bigger than Z_0. We shall look at the implications of this
later. We shall also terminate to an effective voltage of +3v.

Thus we have

$$V_T = 3 = 5\left(\frac{R_2}{R_1 + R_2}\right)$$

$$R_T = 110 = \frac{R_1 R_2}{R_1 + R_2}$$

from which it follows that R_1 = 180 Ω and R_2 = 270 Ω.

We must now calculate the current which the open-collector gate
will sink. In the low state there is a maximum of 5 v across R_1
which gives a current of 28 mA. Since there are two resistors, one at
each end, the total current is almost 60 mA. This is quite a high
current, although there are devices available which can supply it.
It is also quite permissible to use two lower-current devices in
parallel.

Clearly, this bus can be driven at any point, and an output
obtained from any point.

We remember that we overterminated the bus by 10 per cent.
This leads to a reflection of $(110 - 100)/(110 + 100)$, that is about
5 per cent, which we can safely ignore. In general a mismatch of x per
cent leads to a reflection (voltage) of ½ x per cent if x is small.

We must now consider the effect on our transmission line model
of connecting other driver and receiver devices to the bus.

Provided the connecting stubs of these other devices are kept short,
less than 100 mm for Schottky TTL or less than 300 mm for standard
TTL, we can consider their effect as capacitive and, in fact, they
merely lower the characteristic impedance of the bus, since
$Z_0 = \sqrt{(L/C)}$ and we are increasing C. The connections should, however,
be kept as short as possible.

In most cases (for example, twisted pair or printed backplane),
Z_0 will be greater than 100 Ω and we shall not require such 'meaty'
drivers as in the example given.

Use of Flat Multiway Cable for Logic Interconnection

Many manufacturers [for example 3M, Ansley Berg (Du Pont)] are
making a flat cable system with crimped connectors, which is in many
ways ideal for bussing signals around a logic system provided some
simple rules are followed.

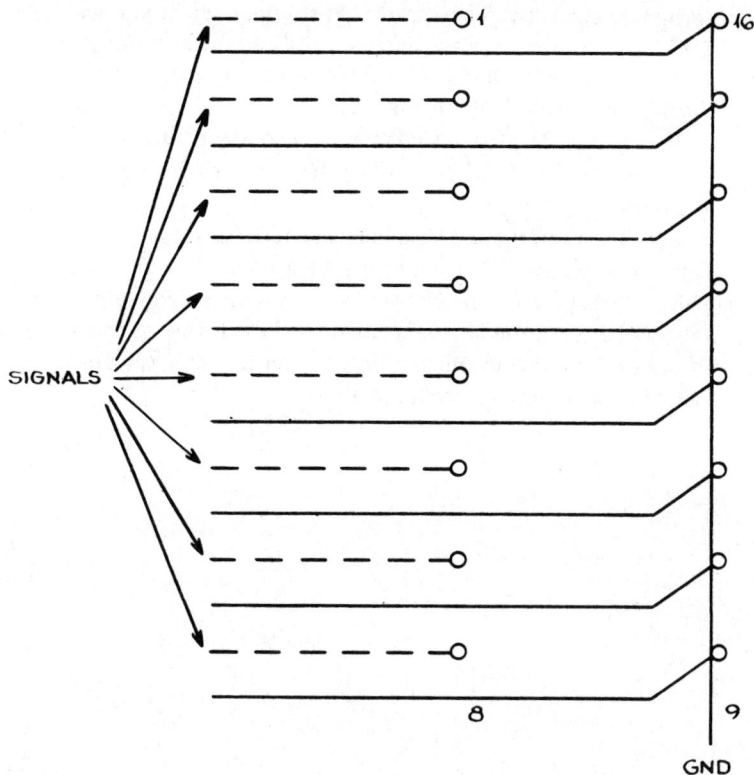

Figure 4.9

(1) Use the cable with alternate conductors grounded. This means that on a 16-way cable you would only send 8 signals. (+5 v would of course do just as well as ground.) If the cable is terminated in a standard 16-pin DIP plug and socket it is simple to carry out this grounding since alternate conductors are brought out to opposite sides of the plug as shown in figure 4.9.

One manufacturer (Ansley) is introducing a cable with conductors on half the pitch (0.025 instead of 0.050 in.). They are also introducing a connector which automatically grounds alternate conductors via a single pin. This new cable will represent a useful advantage in reduced size.

(2) Terminate correctly. There is a slight subtlety which you will notice if you refer back to figure 4.9. All the signal conductors except one have a ground at each side of them — so we are dealing with two signal impedances.

For 'Scotchflex' cable (3M) the edge conductor has an impedance of 140 Ω while those in the centre have an impedance of 110 Ω.

It is suggested that all signal lines be terminated in 140 Ω. This will give about 12.5 per cent reflection on 7 of the lines, which is acceptable. Alternatively a compromise figure about halfway between the two values can be chosen. One of the advantages about transmission lines in logic is that impedance matching need only be approximate.

5

Component Pulse Response

Until digital systems became widespread, particularly high speed ones, steady state sine waves were the normal state of affairs. It is well known that in VHF and microwave regions components cannot be thought of as lumped and specific models must be developed. This attitude is correct and must be applied when attempting to understand the performance of a component in a fast step situation. Firstly, we must have a clear understanding of what a step or pulse is. The leading edge of a step is a shock wave. It is a TEM wavefront which travels at the speed of light. Now it is mathematically possible to take a single step and analyse it using Fourier analysis.

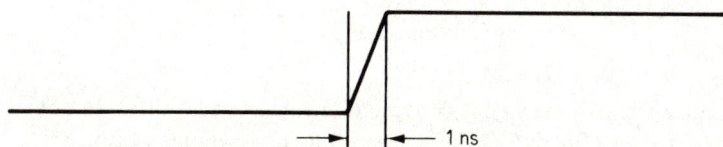

1 ns

Figure 5.1

A single step has a risetime of 1 ns (see figure 5.1); to produce this from sine waves means combining an infinite number which have existed from − infinity to + infinity. Now this can be seen to be quite absurd and not practical at all. Instead, let us take the single event in figure 5.1. A 1 ns step produced from Schottky TTL or ECL (Fairchild now do F100K Series with 700 ps risetimes). It is important to realise that as this wavefront travels along a transmission line everything in front of it has no knowledge of its existence whatsoever. There can be no action at a distance. This is because the effects of

27

any event take a finite time to propagate outwards from the source
of the disturbance.

Any component, be it a resistor or a capacitor, for example, is
distributed throughout space. It is therefore convenient to think of
these components as transmission lines, especially when considering
their performance in a high speed digital environment.

Resistor

It is well known that terminating 50 Ω co-axial cable in a resistor of
the same value is called 'perfect termination'. There is no reflection
of the step. Therefore we can deduce that the signal cannot
differentiate between a transmission line of characteristic impedance
50 Ω and a 50 Ω resistor. In fact the resistor can be clearly seen to be
a transmission line of two distinctly separate portions (see figure 5.2).

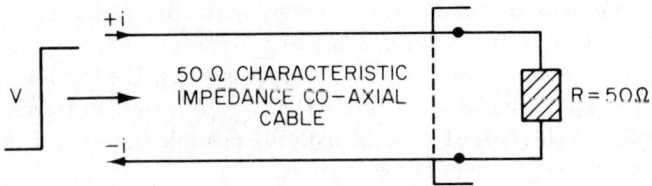

Figure 5.2

The energy in the step is converted into heat and is diffused out
into space (see figure 5.3). Of course a purist viewpoint would say

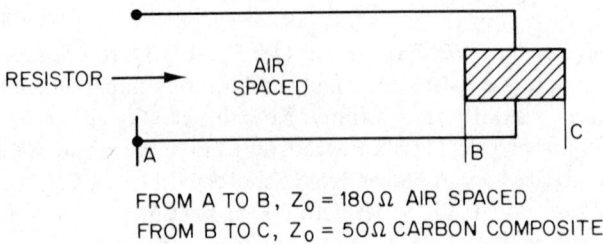

FROM A TO B, $Z_0 = 180\,\Omega$ AIR SPACED
FROM B TO C, $Z_0 = 50\,\Omega$ CARBON COMPOSITE

Figure 5.3

that there is a small discontinuity, but in fact this is so short —
about 12 mm — we can treat the component as an ideal one. However,
the engineer must always bear in mind that it is a very short transmission line before he makes the decision to neglect its effects. This
is sometimes referred to as the series inductance of the resistor.

Capacitor

Firstly, take the case of a parallel-plate capacitor (see figure 5.4). A fast
step propagates down a transmission line with a capacitor connected
to the far end. The step arrives at the end of the transmission line and

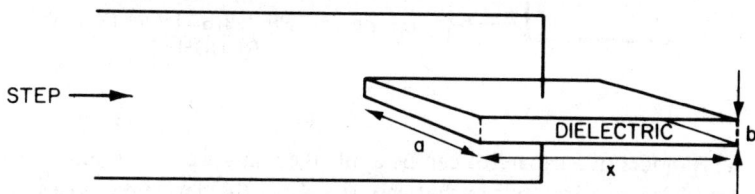

Figure 5.4

sees a very low characteristic impedance compared with its own Z_0 of
50 Ω, say. This is just like a short circuit and so most of the signal
is reflected and only a small portion of it enters the parallel plates
of the capacitor. This small amplitude travels down the parallel plates
and comes to an open circuit at the end. Voltage doubling takes place
due to a reflection coefficient of +1. At the other end of the capacitor
it sees a characteristic impedance much greater than its own
($Z_{0C} = 0.1\ \Omega$, $Z_{0T} = 50\ \Omega$) and thus the voltage step is reflected
once again. The process carries on until the capacitor becomes charged
up to the supply voltage. We have assumed that the cable connected
to the capacitor is long compared with the actual length of the
component, therefore we can ignore reflections in the cable.

The key parameters for the pulse response of a capacitor are
therefore

(1) characteristic impedance

$$Z_0 = \sqrt{\left(\frac{\mu}{\epsilon}\right)} \frac{b}{a}$$

(2) time delay

$$T_d = \sqrt{(\mu\epsilon)}x \text{ s}$$

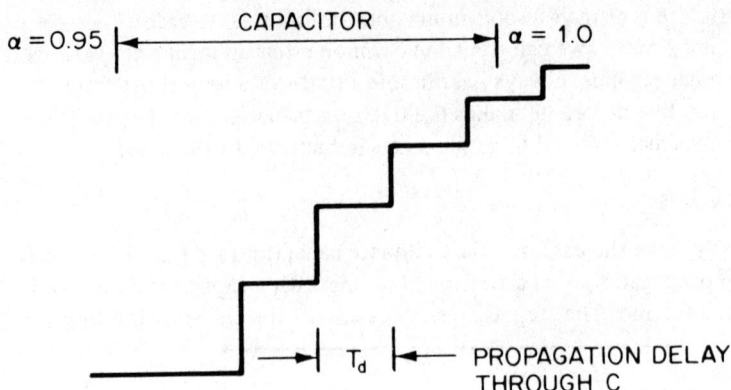

Figure 5.5

This model of a capacitor can be applied equally well to circular plates (disc ceramics), but in that case the Z_0 of the device decreases as the step travels out from the centre of the two plates (see figure 5.6).

Now, the established equation for the charging of a capacitor from a step input is

$$V = iR + \frac{1}{C}\int i \, dt$$

Looking at the second term we see that the constant C is the total capacitance of the component and so it is not valid in a very fast step situation. This is because at the instant the step meets the capacitor the voltage developed across the plates initially cannot be defined by a value which is a function of the total size of the device. The step has no knowledge of the actual length of the plates until it reaches the far end and sees an open circuit.

Even faster edges can be produced by using a time domain reflectometer (TDR). This produces a 25 ps risetime edge which enables clear waveforms to be seen for standard components. In 50 ps, for example, a wavefront will travel about 10 mm in air.

Therefore when components are being used in a digital environment it is of no value to use stated parameters which have been based on sine wave measurement and calculation. To have a clear understanding of how a component behaves in a fast step response situation engineers must use the transmission line parameters developed above.

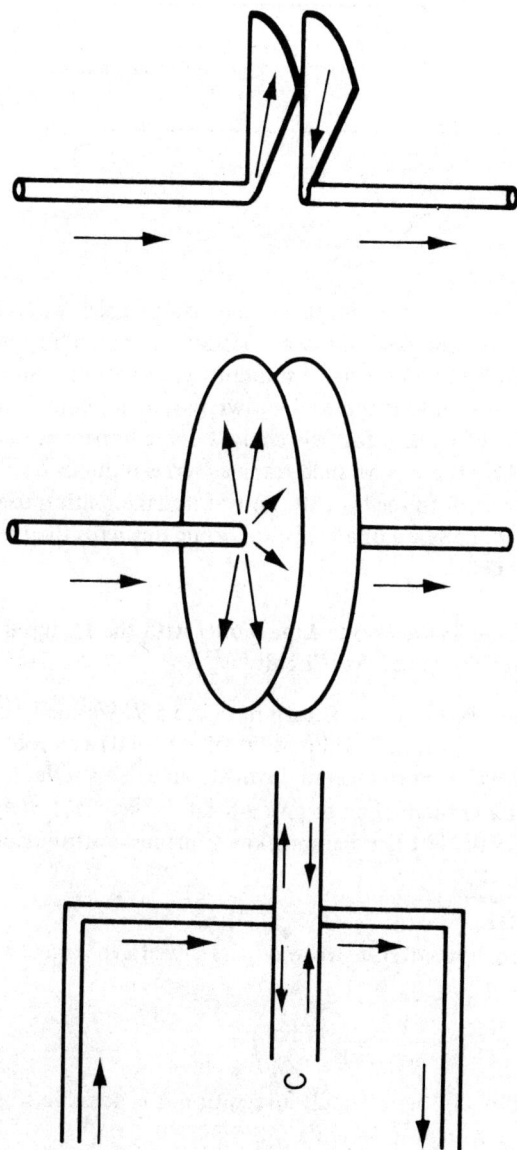

Figure 5.6

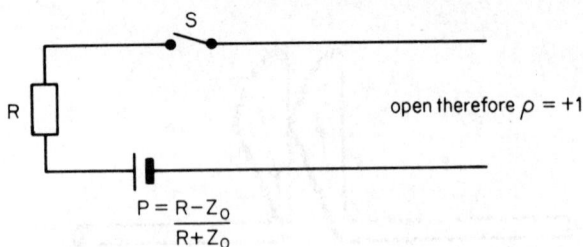

$$P = \frac{R-Z_0}{R+Z_0}$$

open therefore $\rho = +1$

Figure 5.7

Even in a sinusoidal situation this model can be used; when the wavelength of the impressed sine wave becomes comparable with the length of the leads and the plates a standing wave pattern will be present. It can be seen that the well known parameter equivalent series resistance (ESR) is in fact the characteristic impedance (Z_0) of the device. Also there is no such thing as series inductance.

It would be useful to see how the conventional equation used for describing the charging up of a capacitor compares with the transmission line model.

Comparison of the Transmission Line Model with the Lumped Model for a Capacitor in an *RC* Circuit

Consider a transmission line as shown in figure 5.7. We shall assume that $R \gg Z_0$. When switch S is closed (at time $t = 0$) a step of voltage $VZ_0/(R+Z_0)$ is propagated down the line. This reflects from the open at the right-hand end to give a total voltage $2VZ_0/(R+Z_0)$. Reflection from the left-hand end makes a further contribution of

$$\frac{VZ_0}{R+Z_0} \frac{R-Z_0}{R+Z_0}$$

and so on. In general, after n two-way passes we have (where voltage after n passes = V_n)

$$V_n + 1 = V_n + \frac{2VZ_0}{R+Z_0} \left(\frac{R-Z_0}{R+Z_0} \right) n \tag{5.1}$$

In order to avoid a rather difficult integration it is possible to sum this series to n terms using the formula

$$\frac{a(1 - v^n)}{1 - v} \tag{5.2}$$

where a is the first term of a geometrical progression and v the ratio between terms. (This formula is easily verified by induction.)

Substituting the parameters from equation 5.1 into equation 5.2

$$a = \frac{2VZ_0}{R+Z_0} \tag{5.3}$$

$$v = \frac{R-Z_0}{R+Z_0} \tag{5.4}$$

we obtain

$$V_n = \frac{\frac{2VZ_0}{R+Z_0}\left[1 - \left(\frac{R-Z_0}{R+Z_0}\right)n\right]}{1 - \frac{R-Z_0}{R+Z_0}} \tag{5.5}$$

$$= V\left[1 - \left(\frac{R-Z_0}{R+Z_0}\right)n\right] \tag{5.6}$$

Note This is the correct description of what is happening as a capacitor charges. We can now go on to show that it is approximated by an exponential.

We have

$$V_n = V\left[1 - \left(\frac{R-Z_0}{R+Z_0}\right)n\right] \tag{5.7}$$

Consider the term

$$T = \left(\frac{R-Z_0}{R+Z_0}\right)n$$

$$= \left(\frac{1 - Z_0/R}{1 + Z_0/R}\right)n$$

If $Z_0/R \ll 1$, this term is asymptotically equal to

$$\left(1 - \frac{2Z_0}{R}\right)n$$

Now define k to be $2Z_0 n/R$. Substitution gives

$$T = \left(1 - \frac{k}{n}\right)n$$

By definition, as $n \to \infty$ we have

$$T = e^{-k} = \frac{e^{-2Z_0 n}}{R}$$

And therefore

$$V_n = V \left(1 - \frac{e^{-2Z_0 n}}{R} \right)$$

Now, after time t

$$n = \frac{V_c t}{2l}$$

Where V_c = velocity of propagation. Therefore

$$V_{(t)} = V \left(1 - \frac{e^{-V_c t}}{l} \times \frac{Z_0}{R} \right)$$

For any transmission line it can be shown that

$$Z_0 = P \sqrt{\left(\frac{\mu}{\epsilon} \right)}$$

$$V_c = \frac{1}{\sqrt{(\mu \epsilon)}}$$

$$C_1 = \frac{\epsilon}{P}$$

where C_1 = capacitance per unit length, and P is the same geometrical factor in each case. The 'total capacitance' of length l of line = lC_1 = C. Hence

$$\frac{V_c Z_0}{lR} = \frac{1}{RC}$$

and therefore

$$V(t) = V(1 - e^{-t/RC})$$

which is the standard result. These results are plotted in figure 5.8.

Further Reading

I. Catt, D. Walton and M. Davidson, 'History of Displacement Current', *Physics Education* (1979).

I. Catt, D. Walton and M. Davidson, 'Displacement Current', *Wireless World* (1978).

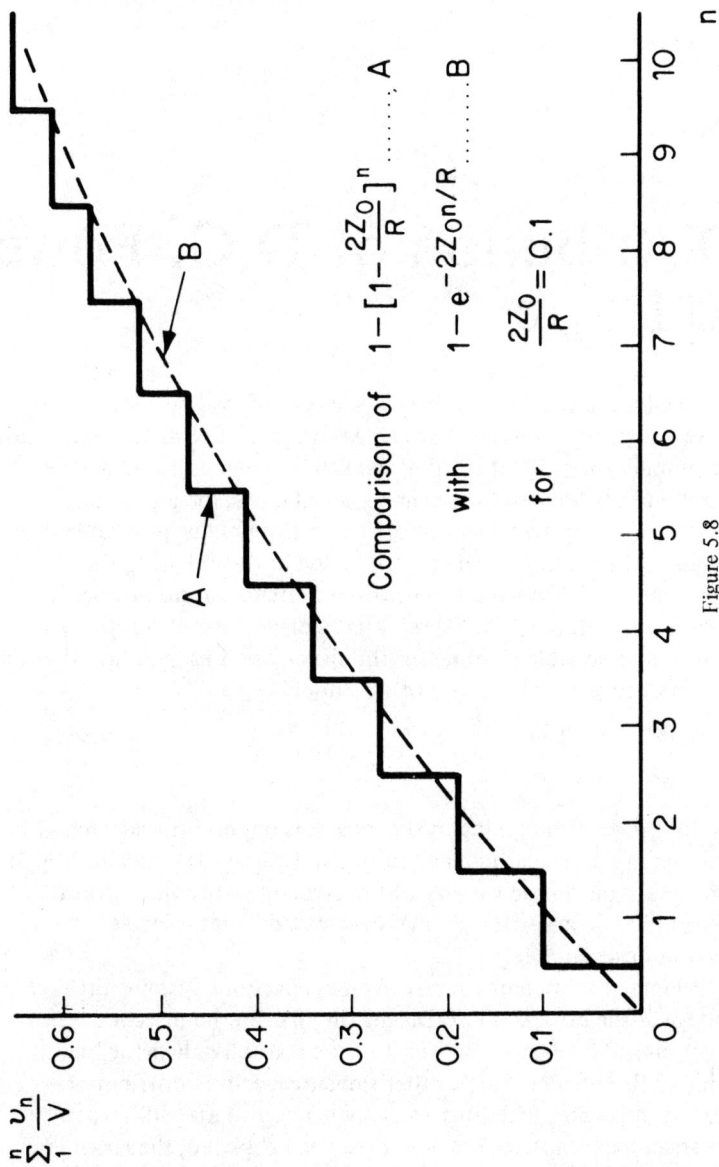

Comparison of $1 - \left[1 - \dfrac{2Z_0}{R}\right]^n$ ....A

with $1 - e^{-2Z_0 n/R}$ ....B

for $\dfrac{2Z_0}{R} = 0.1$

Figure 5.8

6

Distribution of D.C. Power to Logic

In a digital system, sudden massive changes of load current in the power supply serving the logic can easily occur. For instance, it is not uncommon for a 32-bit word of data to be gated on to 32 parallel signal lines. If the word happens to be 'all ones', the amount of current suddenly switched out of the +5 V supply is quite large. For instance, if the lines are shunt terminated in their characteristic impedance of 100 Ω, the total current switched, at 50 mA per line for a 5 V signal, is 1.6 A. We see what a massive event this is when we assume a reasonable risetime for the signal, say 5 ns, and find that the rate of change of current out of the supply

$$\frac{\mathrm{d}i}{\mathrm{d}t} = \frac{1.6}{5 \times 10^{-9}} \text{ A/s} = 320 \times 10^6 \text{ A/s}$$

For proper functioning of the logic it is important that such a massive, rapid current demand should not cause a significant drop in the voltage bus in the vicinity of the demand at any time after that demand has occurred. (A 50 mV drop could be regarded as the maximum acceptable.)

A hierarchy of energy reservoirs, or capacitors, sustains the d.c. voltage at the critical point. During the first few nanoseconds, a small capacitor, near to the point of the suddenly changing load, supplies the current. Later, other similar capacitors distributed around the logic a little further away help out. Later still, when all the small local capacitors would have been depleted, the larger energy reservoir of, say, 20 000 μF in the power supply itself sustains the voltage.

36

As this final reservoir begins to be depleted, the series pass transistors in the power supply alter their current levels to supply the deficiency.

Figure 6.1 shows a typical hierarchy of energy stores such as might be found in a system in use today. At every stage it is deficient. Starting at the right, each capacitor has to have become seriously depleted, its voltage dropping by a volt or two (compared with the maximum reasonable drop of 50 mV) before there is enough voltage and time developed across the next inductor to the left to cause it to pass the required increase of 1.6 A. The situation is unsatisfactory by orders of magnitude, showing that it is not merely some systems that are deficient. In fact, most systems in service today are unable to supply a perfectly normal change in current demand at one point in the system while keeping the d.c. voltage bus near to its nominal value.

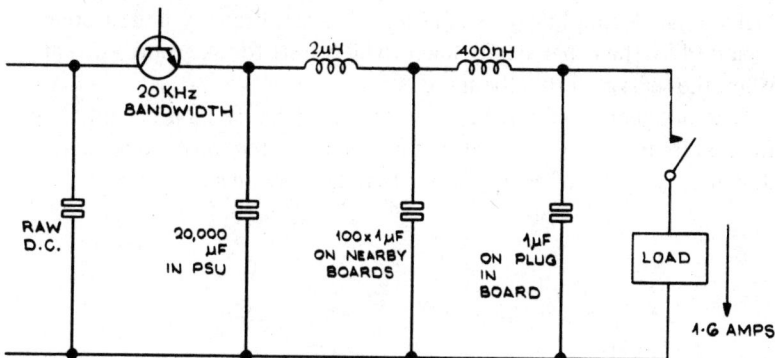

Figure 6.1

Why do systems continue to function if the above assertions are true? The answer is that an important element in the total picture was omitted from the diagram, and that was the other steady loads. When the 1.6 A is switched on, the d.c. voltage supply begins to sag. The result is that, with a reduced voltage across them, the other circuits demand less current. If the d.c. voltage supply drops by 5 per cent, the current taken by an integrated circuit falls by even more than 5 per cent. (The current drop is more than linear with voltage because some of the available voltage is 'lost' in fixed transistor V_{be} drops. This means that if there is a 5 per cent drop in the full voltage, the percentage drop across resistors in the circuit is more than 5 per cent.)

If the full system load is 30 A, a further load of 1.6 A only calls for the removal of about 5 per cent from this standing load, which is probably achieved by a drop of only some 3 per cent in the d.c. voltage supply; a mere 150 mV in the case of a 5 V supply. So we see that in a traditionally designed system, the voltage supply decoupling capacitors are to quite a degree merely a gesture of intent, and the real voltage decoupling is by the steady loads. This is unsatisfactory, because any change in d.c. voltage levels leads to a degradation of performance and possible failure. Timing circuits are upset, and margins against other types of noise are reduced.

When the 1.6 A load is suddenly switched off, a similar problem arises. This time, the voltage bus rises above its correct value in order to repel the now unwanted current which continues to arrive. The amount of over-voltage developed is about equal to the amount of under-voltage which occurred in the previous case. As before, the real voltage decoupling (or stabilisation) in traditionally built systems is caused by the other steady loads, which start to take more current when the voltage across them increases.

The designer should make a conservative (that is, large) estimate of the maximum change of load that can be expected in the logic, and design a hierarchy of energy stores that can sustain such a load.

7

Local Decoupling of Voltage Supplies by Printed Circuit Voltage Planes

When a logic gate switches on, a signal is launched down the signal line, the current step i being taken from the positive supply line, and an equal and opposite current $-i$ being dumped into the 0 V line at the point where the logic signal originated (see figure 7.1).

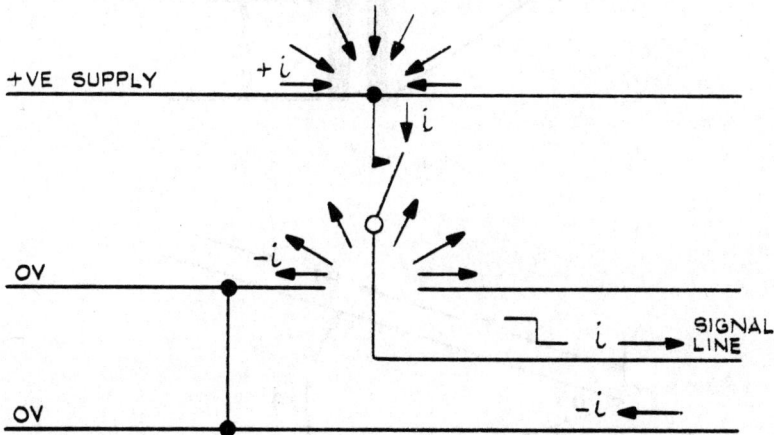

Figure 7.1

As a result, in order to satisfy Kirchhoff's law that the total current at any point must be zero, a circular wave of current and voltage — like the ripple caused by a stone dropped into a pond — flows out between the positive supply voltage plane and the 0 V plane. It is important that the impedance seen by such a signal flowing out between the voltage planes, which can be regarded as a source impedance of the logic signal, should be small. Otherwise the positive voltage difference between the two planes at the signal source will temporarily collapse.

The Pie-shaped Transmission Line

In order to understand the nature of the decoupling action at a point between parallel voltage planes, it is easiest to approach it by way of the parallel-plate transmission line and the pie-shaped transmission line (see figures 7.2 and 7.3).

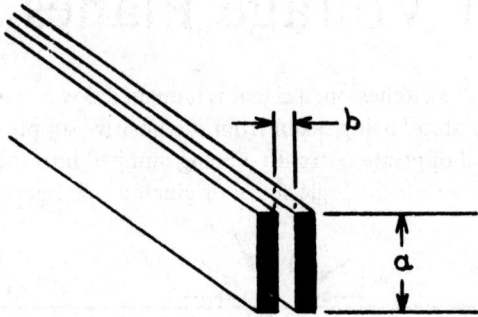

Figure 7.2

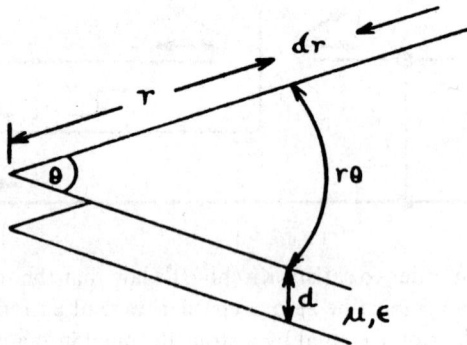

Figure 7.3

The formula for the characteristic impedance of a parallel-plate transmission line of width a and separation b with negligible fringing at the edge (because a is much bigger than b) is

$$Z_0 = \frac{b}{a} \sqrt{\left(\frac{\mu}{\epsilon}\right)}$$

This formula still applies for each small section of a transmission line where the width a is varying, in particular for a pie-shaped transmission line (figure 7.3), where over a distance δr it becomes

$$Z_0 = \frac{d}{r\theta} \sqrt{\left(\frac{\mu}{\epsilon}\right)}$$

Now if $\theta = 2\pi r$, that is, we are considering a complete plane (figure 7.4), we get

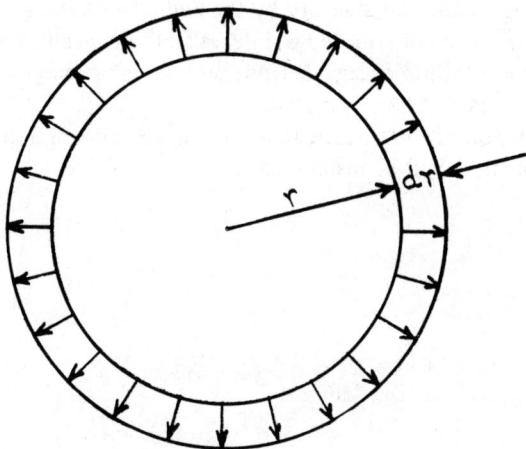

Figure 7.4

$$Z_0 = \frac{d}{2\pi r} \sqrt{\left(\frac{\mu}{\epsilon}\right)}$$

Now we know that, in a medium with permittivity ϵ and permeability μ, the outwards velocity of the signal must be $1/\sqrt{(\mu\epsilon)}$, therefore the velocity is

$$c = \frac{r}{t} = \frac{1}{\sqrt{(\mu\epsilon)}}$$

where time t is the time since the signal was introduced at the centre. So using this last equation to substitute for the distance r we get

$$Z_0 = \frac{d\mu}{2\pi t} \ \Omega$$

where d is in metres.

A reflection related to Z_0 arrives back at the centre at time $2t$. If Z_0 is small when $2t = t_r$ (the risetime of the output of the logic gate), then natural decoupling between planes is satisfactory.

As an example, if $2t = 1$ ns, $d = 0.5$ mm, $\mu = 4\pi \times 10^{-7}$

$$Z_0 = \frac{0.5 \times 10^{-3} \times 4\pi \times 10^{-7}}{2\pi \times \frac{1}{2} \times 10^{-9}} = 0.2 \ \Omega$$

This calculation shows that it is possible to keep the decoupling between voltage planes satisfactory by the addition of extra decoupling capacitors distributed at intervals of a few centimetres, to prevent superposition of signals from generating unacceptably large voltage transients between planes.

The proportion of the surface area occupied by the miniature tantalum capacitors will be insignificant.

8

Printed Circuit Board Layout for High Speed Schottky TTL

This subject is not a difficult one; indeed, the mathematics used is extremely elementary. The problems are caused rather by the historical progression from analog to digital techniques, with the consequent carrying over of well tried analog techniques into the digital environment. Unfortunately, the requirements for digital circuitry are frequently opposite to those needed by the analog variety, and hence there is a need for a complete reconsideration of the requirements.

Low Inductance Bussing

To understand the criteria which determine how the +5 V and 0 V lines should be distributed to the TTL, first take the case of a TTL gate driving its output line from low to high. For the gate to drive the output line high it must pass current into it. The output line must be considered as a transmission line of impedance Z_0 if its length exceeds 100 mm. In practice, Z_0 will be in the region of 100 Ω, and for a single logic signal changing from low to high the instantaneous output current will be given by $I_0 = 5/100 = 50$ mA. This current must be obtained from the supply rails in a time comparable to the risetime of the signal. If, for Schottky TTL, $t_{r(min)} \approx 1.5$ ns, then charge must be transferred from the decoupling capacitor to the gate and hence to the output line in this time. Remember that charge is obstructed

43

from flowing into the gate by the inductance, L, of the loop **ABCD** in figure 8.1. If this is approximately 20 mm square with reasonable track width, then, using the formula for parallel wires

$$L = \frac{\mu_0}{4\pi} \ln \frac{a}{r} \text{ H/m} \approx 30 \text{ nH}$$

The e.m.f. dropped across L will then be given by $E = -L \, di/dt$. Therefore

$$E = \frac{30 \times 10^{-9} \times 50 \times 10^{-3}}{1.5 \times 10^{-9}} = 1 \text{ V}$$

This is a considerable voltage, and it should be remembered that it is the result of a single gate switching. If all four gates in a pack switch together, the currents will be additive, and the rail will fall by 4 V.

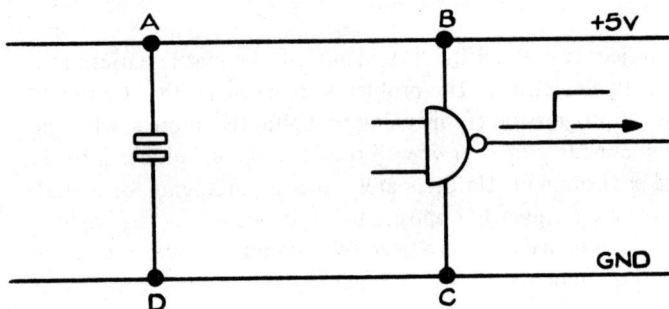

Figure 8.1

The first requirement of a power distribution system must therefore be low inductance between the IC and the decoupling capacitor. This is achieved by the track layout shown in figure 8.2, where a low inductance path from C to the IC is provided by keeping the +5 V and 0 V tracks close together.

Manufacturers of ICs usually specify one decoupling capacitor for every 5 to 10 ICs, which, with the track layout shown in figure 8.2a, results in prohibitively high inductance between the capacitor and the worst-case positioned IC. The safest course is to provide the track layout as in figure 8.2b, but also to put one capacitor adjacent to each IC. Clearly, this can be achieved by having one capacitor for each pair of ICs.

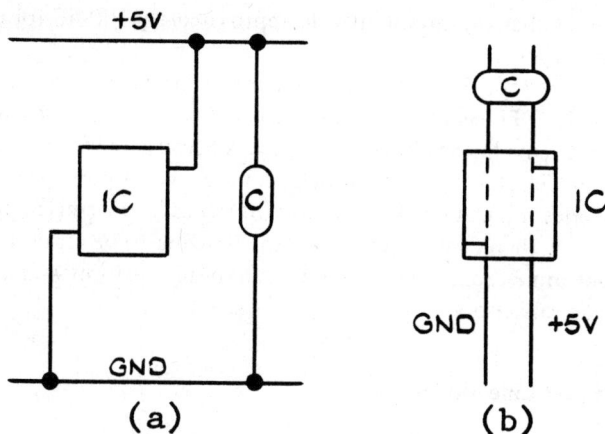

Figure 8.2

Decoupling Capacitors

The foregoing argument shows that the capacitor is better thought of as a reservoir capacitor which supplies the local, instantaneous current. Some manufacturers specify capacitors for IC decoupling by giving the maximum pulse risetime, which corresponds to a maximum current for a given size of capacitor. For instance, a 47 nF capacitor specified at 50 V/μs can supply a current given by

$$i = C\frac{dv}{dt} = 47 \times 10^{-9} \times \frac{50}{10^{-6}} = 2.5 \text{ A}$$

which is adequate in the context of the previous calculation.

The other check to make is that the current drawn from the capacitor does not cause its voltage and hence the rail voltage to fall excessively. If the local demand is equal to 10 gates switching, the current demand will be 500 mA; to be safe, assume that this demand lasts for 10 ns, and design for a voltage drop at the capacitor of 50 mV. Thus

$$i = C\frac{dv}{dt}$$

$$0.5 = C\frac{50 \times 10^{-3}}{10 \times 10^{-9}}$$

$$C = 100 \text{ nF}$$

This suggests that we should provide approximately 100 nF for each pair of packages.

It might be thought that radio frequency type capacitors are necessary for TTL decoupling, but this is not so. To show why will be explained in a later volume. Briefly, it is because the frequently adopted model of a capacitor, which proposes that it possesses a lumped series inductance, breaks down in the case of a single applied step. There is therefore no reason for the designer to be afraid to use non-ceramic capacitors. 1 μF tantalum beads perform well as decoupling capacitors.

Transmission Line Model

The best way to think of the power distribution system is as a transmission line, with each package connected to an ideal voltage source via an impedance equal to the transmission line impedance. (A package at the centre of a power bus will see two transmission lines in parallel and hence half the impedance. We shall adopt the worse figure for the purpose of this argument.) This impedance must be sufficiently low for negligible voltage transients to be produced on the line by gates switching within the package. The impedance of a transmission line is given by $Z_0 = \sqrt{(L/C)}$ where L and C are the inductance and capacitance per unit length respectively. To calculate Z_0 for the case of two tracks close together

$$L = \frac{\mu_0}{4\pi} \ln \frac{a}{r}$$

where μ_0 is 5, a and r are taken as 2 mm and 0.5 mm. Therefore

$$L = 0.6 \, \mu\text{H/m}$$

If a 100 nF capacitor is placed every 50 mm along this line, then

$$C = 100 \times 20 \, \text{nF/m} = 2 \, \mu\text{F/m}$$

therefore

$$Z \approx 0.5 \, \Omega$$

An instantaneous current demand of 200 mA — corresponding to 4 gates switching — will produce a voltage transient of 100 mV. This is only just acceptable, and suggests that the value of C should be increased. Note, however, that laying out the tracks with wider spacing and using smaller capacitors — 10 nF for every few ICs, which is not uncommon — will create a situation much worse than this.

Auto-decoupling in TTL

In the context of the preceding remarks, some readers may wonder how systems which they have seen or have worked with managed to function at all, since it is common to see most or all of the above design guidelines violated. To see the answer to this paradox, consider the structure of the TTL gate output circuit, when this is driving the gate input low shown in figure 8.3.

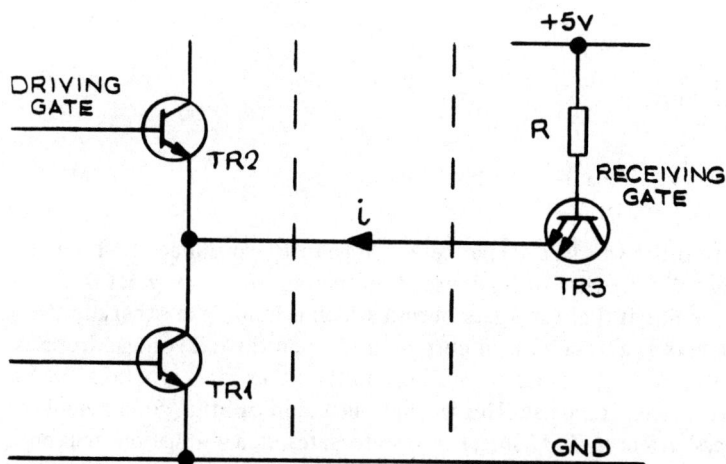

Figure 8.3

According to the specification for, say, a 7400, the typical values of i and R are 1.0 mA and 4 kΩ respectively. When the gate output is low it sinks a current i, given by

$$i = \frac{V_{cc} - V_{be} - V_{ce_{sat}}}{R}$$

where V_{be} is the base–emitter voltage of TR3 and $V_{ce_{sat}}$ is the collector saturation voltage of TR1.

If V_{be} and $V_{ce_{sat}}$ = 0.7 V, to take a worst-case example, and V_{cc} = 5 V, then

$$i = \frac{3.6}{R}$$

Now consider what happens if the rail voltage drops, due to a transient load imposed by the output of another gate switching. When V_{cc} drops, there is (to a good approximation) no change in the V_{be} drops. Suppose the rail drops by 10 per cent. Then

$$i_1 = \frac{5 - 1.4}{R}$$

$$i_2 = \frac{4.5 - 1.4}{R}$$

Therefore

$$\frac{i_1 - i_2}{i_1} = \frac{0.5}{3.6} = 14\%$$

In other words, a 10 per cent change in V_{cc} produces a 14 per cent change in the current load placed on the rail. In effect, what is happening is that each gate output which is holding another input low acts as a 'reservoir' of current, and when the rail voltage drops as another gate drives its output high all the other gates give up some of their current to assist. This is what one could call the 'good neighbourliness' effect in TTL. In general, some gates on a voltage bus will be low and so act as current supplies. The problem arises when none or only a few are in this state — a critical situation for a badly designed system and one which could cause failure. It should be remembered that a logic system should work for all possible combinations of states which can occur in practice, and a hazard of this type could have serious consequences. It is therefore insufficient to demonstrate that a system 'works', because if the power distribution system is badly designed there is always the chance of an untested situation bringing about a failure of the system. It is assumed that in a logic system of reasonable size it is impossible to test all possible combinational situations, and doubly impossible to test all possible changes of situation!

The problem with Schottky TTL is that the increase in speed does not allow time for the 'good neighbourliness effect' to act; consequently one is many times worse off with Schottky than with ordinary TTL. Schottky is a less forgiving family than conventional TTL, and much more care must therefore be taken with power distribution to ensure reliable performance.

The Current Spike

As just described, the main cause of transient current demands in a Schottky TTL system is the initial current surge when a gate switches into its transmission line load. The manufacturers' data overlook the mechanism entirely. There is another cause of transient current demand which results from the 'push–pull' design of the TTL output stage shown in figure 8.4.

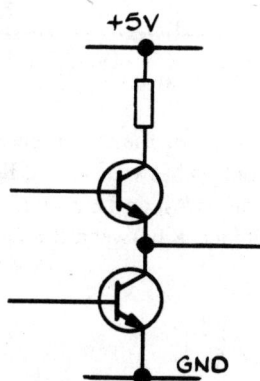

Figure 8.4

The current spike is produced because, on the 0 to 1 transition, the upper transistor turns on while the lower transistor is still turning off. This leads to a current surge of 10 mA with duration of about 10 ns. (See B. Bonham, 'Schottky TTL', T.I. Application Report B93, 1972.) Provided the design guidelines laid down in the earlier sections with regard to power supply bussing and decoupling have been followed, this small additional hazard will be taken care of. In fact, since a logic gate is driving a transmission line which is a resistive rather than a capacitive load, there is no need to provide a totem pole output, and this must be regarded as one of the bad features of the TTL family.

Interconnections

To implement a system successfully using the TTL family it is necessary to interconnect correctly between logic gates.

Transmission Lines   The correct model to use for interconnection between logic gates is a two-wire transmission line. It is impossible to understand how a signal travels from gate to gate without taking the return path into consideration. Indeed it is impossible for a signal to travel without a return path! Consider the two-wire transmission line shown in figure 8.5, in which a zero risetime is propagating to

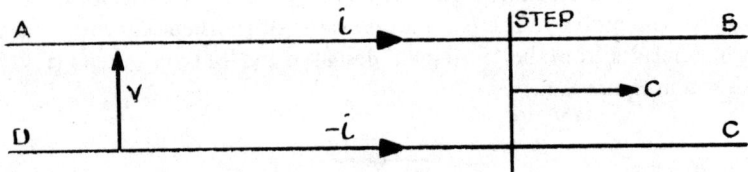

Figure 8.5

the right with velocity c. Ahead of the step there is no current in the wires and no voltage differences between them. Behind the step there is a current i in the direction of AB and a current $-i$ in the direction of DC with a voltage difference V between the wires. It can be shown that $V = iZ_0$, where

$$Z_0 = \sqrt{\left(\frac{L}{C}\right)} = \sqrt{\left(\frac{\mu}{\epsilon}\right)}$$

where Z_0 = characteristic impedance of the line
L = inductance per unit length of the line
C = capacitance per unit length of the line
μ = permeability of medium between the wires
ϵ = permittivity of medium between the wires.

The velocity of propagation c is

$$c = \frac{1}{\sqrt{(LC)}} = \frac{1}{\sqrt{(\mu\epsilon)}}$$

These equations are true for any two-conductor system where the resistance of the conductors can be neglected and the medium between the conductors is well behaved. These conditions are met by tracks

on a printed circuit board for any track width which can be manu-
factured. The step that we have just described is a transverse
electromagnetic disturbance. Since the equation relating current and
voltage on a transmission line is $V = iZ_0$, it follows that the effect of
a transmission line on the driving circuit can be considered in terms
of a resistance $R = Z_0$ connected in place of the line. This was the
procedure followed earlier in calculating the current drawn from the
supply rail by a gate as it switches.

The impedance Z_0 depends on the cross-sectional geometry of the
conductors used, and it is very difficult to calculate except for very
simple cases. It is, however, a relatively slowly varying function of the
geometry [usually logarithmic; see M. A. R. Gunston, *Microwave
Transmission Line Impedance Data* (Van Nostrand Reinhold,
New York, 1972)] and therefore need not worry us too much. For
a track on a printed circuit board laid out according to the design
rules evolved in this paper, a value of Z_0 of around 150 Ω can be
assumed.

One key feature of a board of logic which distinguishes it from
most analog systems is that there are a multiplicity of signal paths
from various points scattered about the board to various other
similar points. It is essential that each of these signal routes has an
adjacent return path. The simplest way, conceptually, to achieve
this is to provide a ground plane on one side of the board. In practice
this is difficult since it usually requires multilayer construction, with
the increased cost and complexity which this entails, in order to
accommodate the signal interconnections. With Schottky TTL it is
not necessary to go to this extreme; all that is required is a ground
grid laid out so that a signal line is never more than 25 mm away from
its return path.

Ground Loops   It might be argued that this scheme leads to ground
loops which, from our experience with analog systems (for example,
audio equipment), are to be avoided. The plain fact is that on a logic
board ground loops are of no importance. The reasons for this are
somewhat complex, but it is probably useful to note one simple
argument. In a high gain amplifier, induction of a few millivolts
at the input due to ground loop pickup can lead to an output of the
same order as the signal. In logic this is not the case; a few millivolts
into a gate input make no difference whatsoever. Hundreds of
millivolts of noise are required before we shall significantly degrade

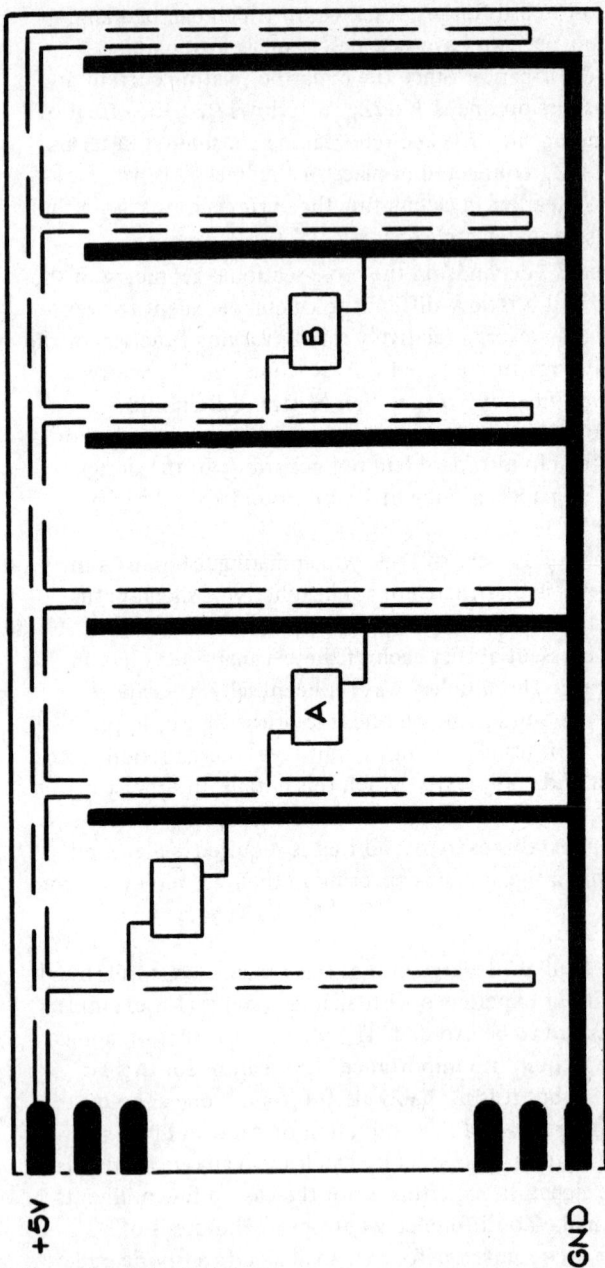

A BAD LAYOUT GIVING HIGH INDUCTANCE AND
FEW ADJACENT SIGNAL RETURN PATHS, WHICH
LEADS TO CROSS – TALK

Figure 8.6

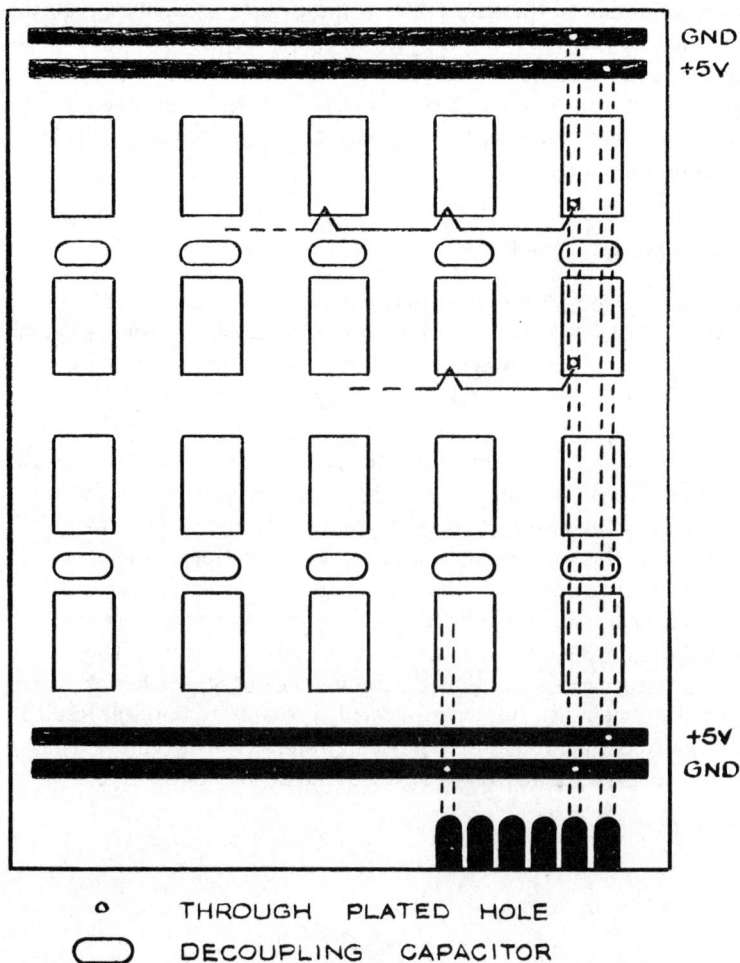

THROUGH PLATED HOLE

DECOUPLING CAPACITOR

Figure 8.7

the noise immunity of a TTL system.

It is probably valuable to examine a situation where a logic board
has been laid out in order to avoid ground loops. A possible layout of
power and ground connections, which is quite common in the industry,
is shown in figure 8.6. Now, if circuit A sends a step to circuit B there
is no adjacent return path. In practice, since a fast step requires a
return path, it will simply use adjacent signal lines as returns, resulting

in the induction of transient noise on these other signal lines. A further consequence is that the input to B will take a longer time to settle with a consequent reduction in the speed of the system. As was explained earlier, the layout in figure 8.6 is also bad from the point of view of placing excessive inductance in the way of charge travelling between ICs and decoupling capacitors.

Recommended Layout

A recommended scheme for laying out a printed circuit board is shown in figure 8.7. The power rails are run as near together as possible along the columns of integrated circuit packages and are interconnected at the top and bottom of the board. These provide return paths for logic signals travelling parallel to them. To provide return paths for signals travelling across the board, the ground pins of the packages are connected together from left to right. (Also consider connecting the +5 V pins of the packages together from left to right.) This track, of the same thickness used for signal interconnections, can be used for this. A tantalum bead $10 \mu F$ decoupling capacitor is provided between each pair of ICs. Note also that ground connections are brought out at regular intervals across the edge connector. These provide return paths for signals travelling on and off the board.

If all these design rules are followed, a reliable system will result and the consequent savings in servicing and testing will amply repay a little consideration given to board layout at the design stage.

9

Crosstalk

Crosstalk between an active line — one carrying a voltage and current step — and a supposedly passive line nearby places a spurious pulse on the passive line. If the spurious pulse is large enough, it could falsely switch circuits on the passive line. Ultrahigh speed circuits require careful consideration of this problem, since it takes less energy to switch a fast circuit than a slow one.

However, to avoid reflection due to mismatch, the designer was forced to make long interconnections as transmission lines of constant Z_0 (as in the graphs shown in figures 9.6 and 9.7) correctly terminated at their destination.

Therefore, when designing for worst-case crosstalk, we need only consider two such lines running parallel for a long distance.

This case has been studied exhaustively by the authors and they have developed a complete theory. The theory is demonstrated by the oscilloscope waveforms in figures 9.2 to 9.5, taken at the indicated points along the active and passive lines. To simplify the explanation, the lines are very long and are placed closer together than would occur in practice. The lines are on the surface of a board, above a buried ground plane (figure 9.1).

The lowest trace in figure 9.2 shows a very narrow pulse introduced at the front end of the active line. If there were no parallel passive line nearby, this pulse would travel down the active line more or less unchanged, in a mode approximating the TEM mode. However, as the other two traces show, the presence of the passive line caused the original narrow pulse to break up into two similar pulses. These travel at different velocities and the smaller pulse is the faster one.

The small crosstalk pulse at the front end of the passive line (seen in

55

the lowest trace of figure 9.3) also breaks up into two pulses. The smaller, faster pulse is equal and opposite to the smaller, faster pulse on the active line. This pair of equal and opposite pulses is the odd-mode signal, which travels in a differential mode down the active and passive lines.

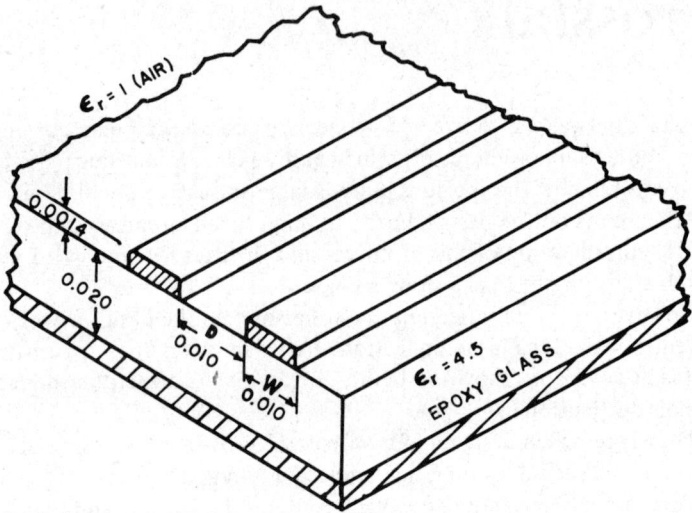

Figure 9.1 Logic board for crosstalk experiments (figures 9.2 to 9.5).

The larger, slower pulse on the passive line is equal to the larger, slower pulse on the active line. This pair of pulses is the even-mode signal, a common-mode signal down both lines with the ground plane as a return path.

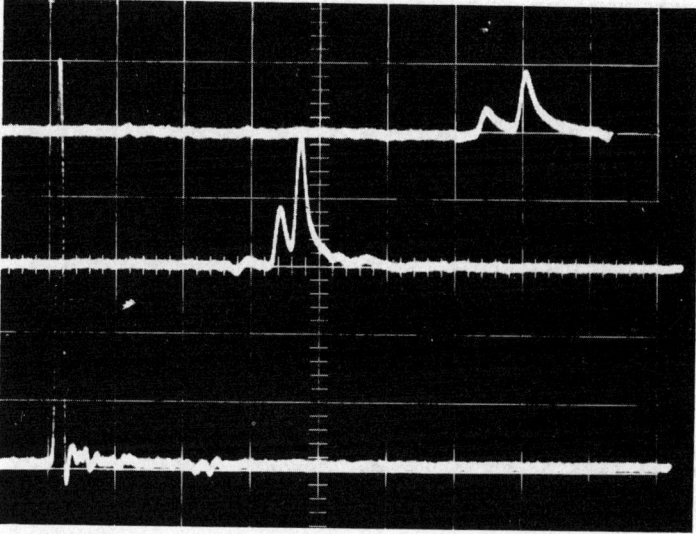

Figure 9.2 Active line. Third trace: front end of line. Second trace: 120 in. down line. First trace: 234 in. down line. Vertical scale 20 mV/div. EH-125 generator. 10 V pulse through 10 dB pad into line. Probing by 500 Ω input of Tektronix 4S2 plug-in unit of 661 oscilloscope. Horizontal scale 5 ns/div.

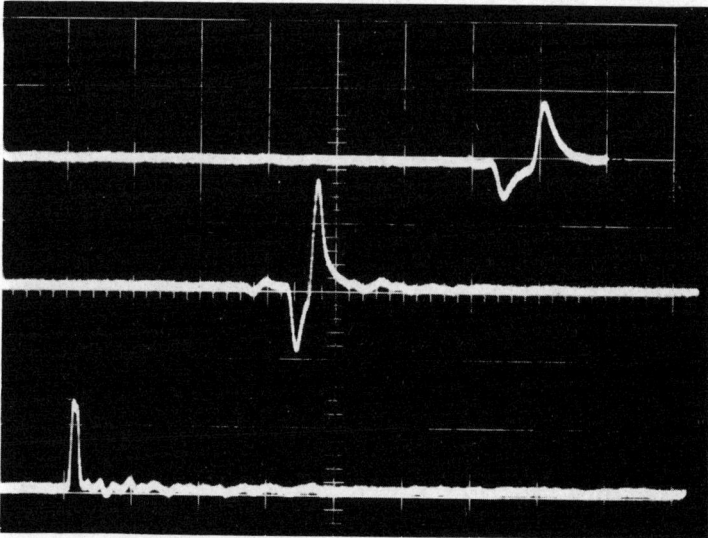

Figure 9.3 Passive line. Third trace: front end of line. Second trace: 120 in. down line. First trace: 234 in. down line.

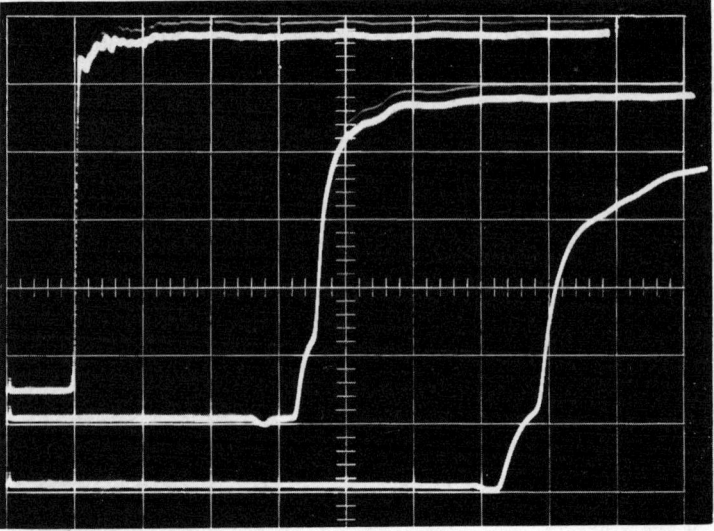

Figure 9.4   Same as figure 9.2 but with a step introduced into the active line instead of a narrow spike.

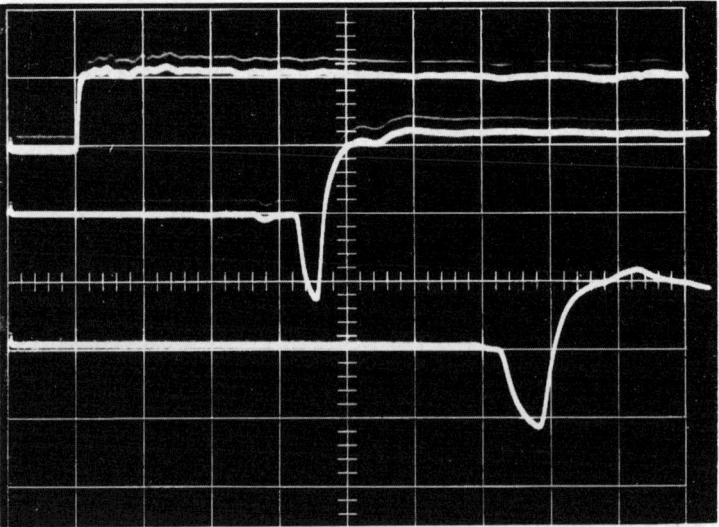

Figure 9.5   Same as figure 9.3 but with a step introduced into the active line instead of a narrow spike.

Step, not Pulse

A narrow pulse is not usually introduced at the front end of an active line in a digital system. Usually, a step representing a transition from the false state to the true state is introduced. The equivalent waveforms for a step are illustrated in the next two oscilloscope diagrams, figures 9.4 and 9.5.

The initial negative-going spike seen in the second and third traces of the passive-line diagram (figure 9.5) is defined as differential crosstalk. It is so named because it results from a velocity differential that causes the odd-mode, or differential-mode, signal to appear on the passive line before the slower even-mode, or common-mode, signal has arrived.

All three traces of the passive-line diagram (figure 9.5) show a positive level equal to the even-mode signal superimposed on the odd-mode signal. This is defined as fast crosstalk; it would appear on the passive line even if the two signals travelled at the same velocity, as they do when conductors are buried between two voltage planes in a multilayer board.

If the lines were short, the full amplitude of fast crosstalk would not have time to appear. Instead of a flat-topped pulse on the passive line there would be a spike of smaller amplitude. This is called slow crosstalk, because the signal risetimes are too slow for fast crosstalk to appear.

Crosstalk Calculations

As initial guidelines, the following may be used.

(1)   If signal risetimes are faster than 2 ns, the designer must consider both fast crosstalk and differential crosstalk.
(2)   If signal risetimes are between 2 and 5 ns, only fast crosstalk need be considered.
(3)   If signal risetimes are slower than 5 ns, slow crosstalk applies.

To calculate differential crosstalk, refer to the graph in figure 9.9. It shows the percentage of velocity difference between parallel surface conductors for the odd-mode signal and the even-mode signal, for 0.010-in.-wide lines.

The maximum amplitude of differential crosstalk equals about 50 per cent of the signal amplitude on the active line. Of course, this

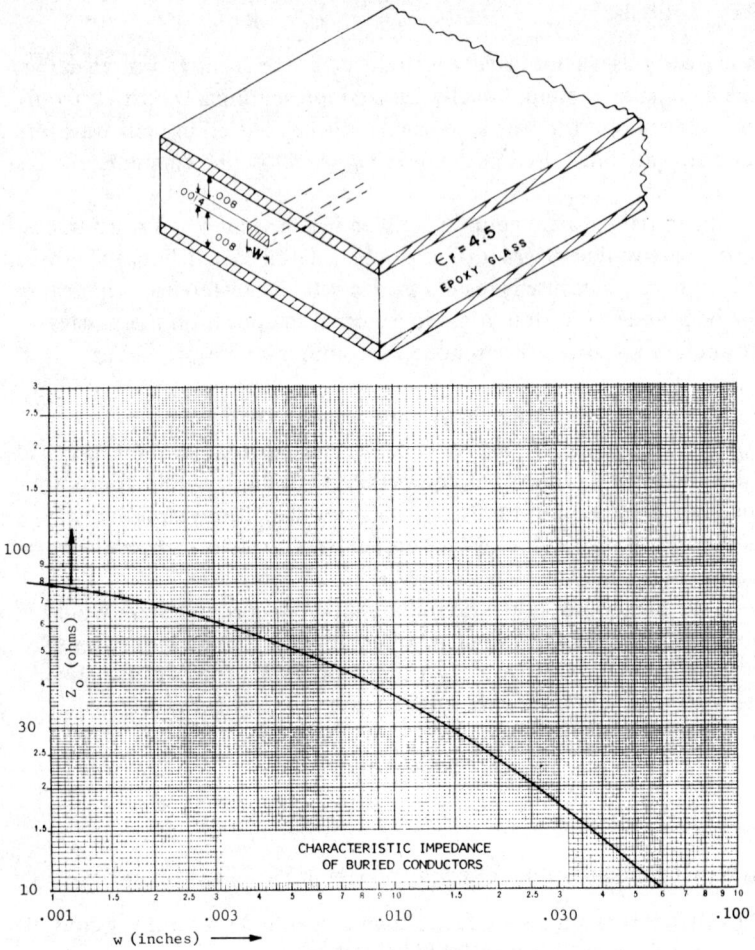

Figure 9.6

amplitude would only be reached in very long lines, such as those used to prepare the oscilloscope diagrams in figures 9.2 and 9.3. In shorter lines of practical length, differential crosstalk can be calculated from the following equation

$$\text{differential crosstalk} = \frac{V_{ag}(\Delta c / C_{avg})Lp}{2C_{avg}t_r}$$

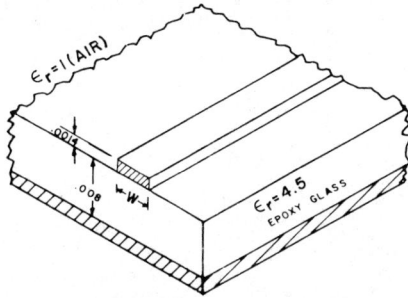

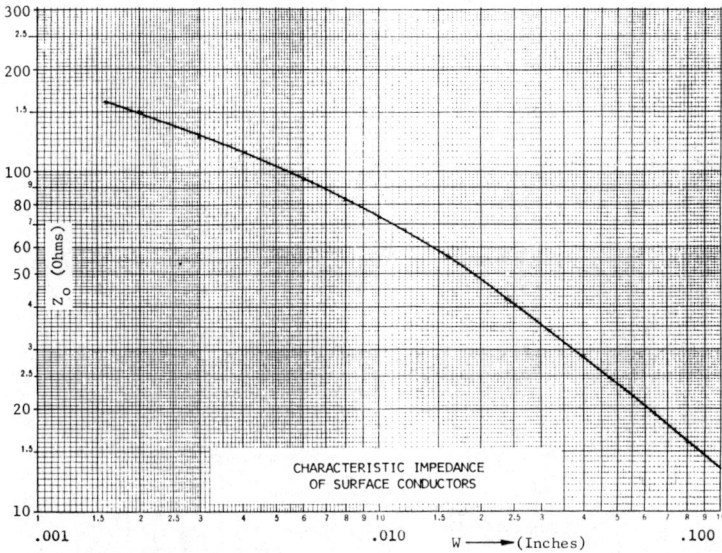

Figure 9.7

where V_{ag} is the signal amplitude on the active line; $\Delta c / C_{avg}$ is the percentage of velocity difference between the differential-mode signal and the common-mode signal; C_{avg} is the average propagation velocity for the two signal modes; L_p is the length of the passive line; and t_r is the risetime of the signal introduced into the active line.

No calculation of fast crosstalk is needed. The two graphs in figures 9.8 and 9.9 give the fast crosstalk for lines of various dimensions.

Slow crosstalk can be determined from the fast crosstalk graphs.

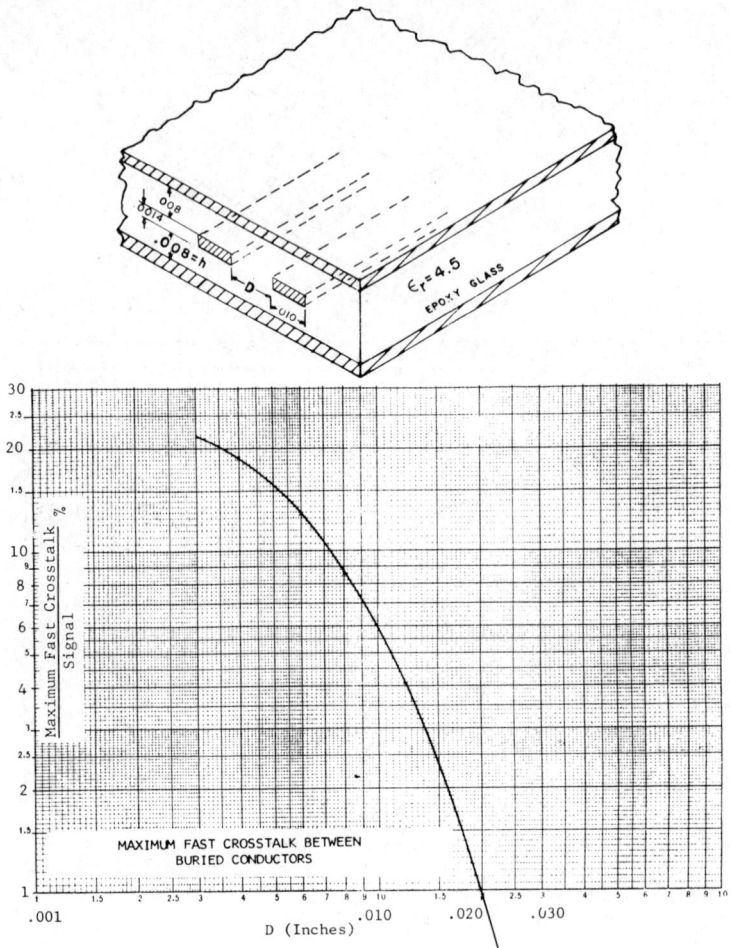

Figure 9.8

Take the fast crosstalk value and reduce that value with the following formula

$$\text{slow crosstalk} = \frac{\text{fast crosstalk} \times 2L_p}{ct_r}$$

where c is signal-propagation velocity, equal here to C_{avg}, the average propagation velocity.

If lines terminate in their characteristic impedance, the amplitude

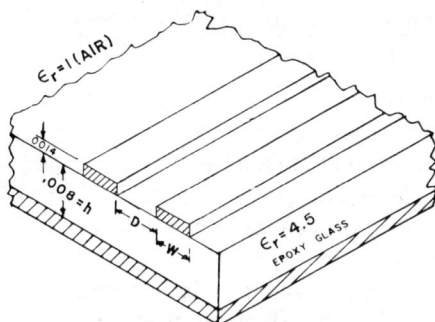

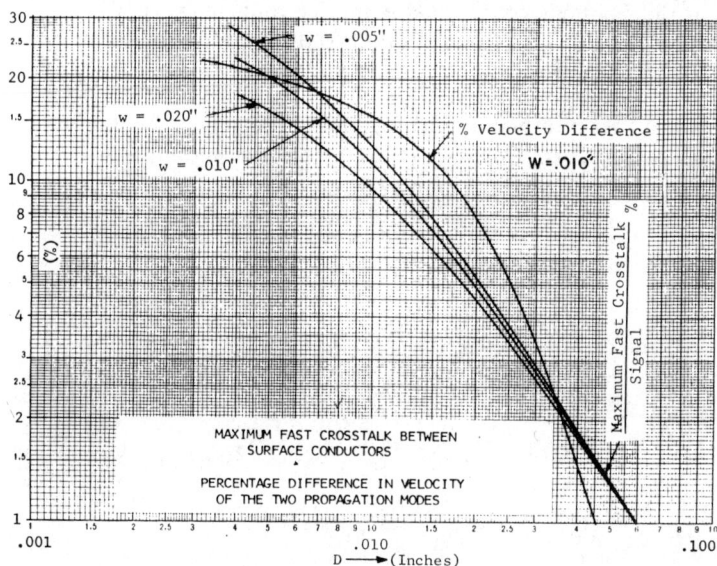

Figure 9.9

of all types of reflected crosstalk is less than the values for initial
crosstalk calculated above, and the designer need not concern himself
with reflected crosstalk.

The ratio in amplitude of the odd-mode signal to the even-mode
signal changes if a resistor is added between ground and the front end
of the passive line. The amplitude of slow and fast crosstalk is reduced
and the amplitude of differential crosstalk is slightly increased. The
changes are not impressive. The designer can keep to the calculations
outlined above when the front end of the passive line is open circuit,

which gives worst-case (maximum) results. Specific examples of what happens are as follows.

(1) If the front end of the passive line is shorted to ground, the first two traces in the upper set of oscilloscope diagrams in figure 9.2 will show spikes of equal amplitude. A slightly increased negative spike of differential crosstalk will result in the second and third traces of the passive line. In the same figure, the amplitude of fast crosstalk will drop to zero. However, no real reduction of fast crosstalk is achieved, because there will still be reflected fast crosstalk to contend with.

(2) If the front end of the passive line is terminated with its characteristic impedance, fast crosstalk will be reduced by about 50 per cent.

For further information on crosstalk between logic interconnections, see Ivor Catt, 'Crosstalk (Noise) in Digital Systems', *I.E.E.E. Trans. Electron. Computers,* EC—16 (1967) 743–63.

10

Energy Current

Oliver Heaviside, who had the advantage of being born later, had a better grasp of electromagnetics than did Faraday or Maxwell, and his view of how a digital signal travels is well worth study.

Whereas the conventional approach to the subject today is to concentrate on the electric current in wires, with some additional consideration of voltages between wires, Heaviside concentrates primarily on what he calls 'energy current', this being the electromagnetic field which travels in the dielectric between the wires. In the quotation below, Heaviside's phrase, 'We reverse this;' points to the great watershed in the history of electromagnetic theory — between the 'ethereals', who with Heaviside believe that the signal is an 'energy current' which travels in the dielectric between the wires, and the 'practical electricians', who like John T. Sprague believe that the signal is an electric current which travels down copper wires, and that if there *is* a 'field' in the space between the wires, this is only a result of what is happening in the conductors.

In his *Electrical Papers*, Vol. 1, (1892) p. 438, Heaviside wrote

'Now in Maxwell's theory there is the potential energy of the displacement produced in the dielectric parts by the electric force, and there is the kinetic or magnetic energy of the magnetic induction due to the magnetic force in all parts of the field, including the conducting parts. They are supposed to be set up by the current in the wire. We reverse this; the current in the wire is set up by the energy transmitted through the medium around it. . .'

The importance of Heaviside's phrase, 'We reverse this;' cannot be overstated for digital designers. It points to the watershed between the 'practical electricians', who have held sway for the last half century, promulgating their theory — which we shall call 'Theory N', the Normal Theory: that the cause is electric currents in wires and electromagnetic fields are merely an effect — and the 'ethereals', who believe what we shall call 'Theory H': that the travelling field is the cause, and electric currents are merely an effect of these fields.

The situation is of course obscured by the many who claim that it is immaterial which causes which. However, experience shows that it is damaging to ignore causality when we are trying to assemble reliable digital systems.

Before continuing with Theory H, we shall quote one early Theory N man, a 'practical electrician' named John T. Sprague. In his book, *Electricity: Its Theory Sources and Applications*, (1892), he ridicules Theory H on p. 239

'A new doctrine is becoming fashionable of late years, devised chiefly in order to bring the now important phenomena of alternating currents under the mathematical system. It is purely imaginary. . . based upon Clerk-Maxwell's electromagnetic theory of light, itself correctly described by a favourable reviewer as "a daring stroke of scientific speculation," alleged to be proved by the very little understood experiments of Hertz, and supported by a host of assumptions and assertions for which no kind of evidence is offered; but its advocates now call it the "orthodox" theory.

'This theory separates the two factors of electricity. . ., and declares that the "current", the material action, is carried by the "so-called conductor" (which according to Dr Lodge contains nothing, not even an impulse, and according to Mr O. Heaviside is to be regarded rather as an obstructor), but the energy leaves the "source" (battery or dynamo) "radiant in exactly the same sense as light is radiant", according to Professor Silvanus P. Thompson, and is carried in space by the ether: that it then "swirls" round (cause for such swirling no one explains) and finds its way to the conductor in which it then produces the current which is apparently merely an agency for clearing the ether of energy which tends to "choke" it, while the conductor serves no other purpose than that of a "waste pipe" to get rid of this energy. . .

'This much, however, is certain; that if the "ether" or medium, or di-electrics carry the energy, the practical electrician must not imagine he can get nature to do his work for him; the ether, &c., play no part whatever in the calculations he has to make; whether copper wire is a conductor or a waste pipe, that is what he has to provide in quantity and quality to do the work; if gutta percha, &c., really carry the energy, he need not trouble about providing *for that purpose*; he must see to it that he provides it according to the belief that it prevents loss of current. In other words, let theoretical mathematicians devise what new theories they please, the practical electrician must work upon the old theory that the conductor does his work and the insulation prevents its being wasted. Ohm's law (based on the old theory) is still his safe guide.

'For this reason I would urge all practical electricians, and all students who desire to gain a clear conception of the actual operations of electricity, to dismiss from their minds the new unproved hypotheses about the ether and the abstract theory of conduction, and to completely master the old, the practical, and common sense theory which links matter and energy together, . . .'

Sprague accurately described Theory N, which has been used in practice by virtually every digital designer, with disastrous results. They must now turn to Theory H to get them out of their difficulties.

In his book *Magnets and Electric Currents* (1898) J. A. Fleming argued on p. 80 for Theory H

'It is important that the student should bear in mind that, although we are accustomed to speak of the current as *flowing in the wire* in one direction or the other, this is a mere form of words. What we call *the current* in the wire is, to a very large extent, a process going on in the space or material outside the wire. Just as we familiarly speak of the sun as rising and setting, when the effect is really due to the rotation of the earth, so the ordinary language we use in speaking about electric currents flowing in conductors retains the form impressed upon it by older and erroneous assumptions as to their nature.'

The reader will have surmised by now that 'energy current', the primary signal which travels down the dielectric from one logic gate to the next, has an amplitude equal to the Poynting vector, $E \times H$.

We shall end this qualitative discussion with some of the more
important quotations from Heaviside, the man who a century ago
brilliantly used the concept of energy current to solve telegraph
problems which closely parallel present-day problems in high speed
digital logic.

In his *Electrical Papers*, Vol. 1, (1892) p. 438, Heaviside wrote

'It becomes important to find the paths along which the
energy is being transmitted. First define the energy-current
at a point to be the amount of energy transferred in unit time
across unit area perpendicular to the direction of transmission. . . .
This is true universally, irrespective of the nature of the medium
as to conductivity, capacity, and permeability, . . . and is true in
transient as well as in steady states. A line of energy-current is
perpendicular to the electric and the magnetic force, and is a
line of pressure. We here give a few general notions.

'Return to our wire from London to Edinburgh with a
steady current from the battery in London. The energy is
poured out of the battery *sideways* into the dielectric at a
steady rate. . . . Most of the energy is transmitted parallel
to the wire nearly . . . . But some of the outer tubes go out
into space to an immense distance. . . . If there is an instrument
in circuit at Edinburgh, it is worked by energy that has
travelled wholly through the dielectric, then finding its way
into the instrument, . . . where it enters . . . and is there
dissipated. . . .

'In a circular circuit, with the battery at one end of a diameter,
its other end is the neutral point; the lines of energy-current
are distributed symmetrically with respect to the diameter.

'On closing the battery circuit (i.e. switching the logic output)
there is an immediate rush of energy into the dielectric. . . . '

11

Grounding in a Digital System

Consider a signal line from A to B, carrying a logic signal which is referenced to the machine frame at each end (see figure 11.1).

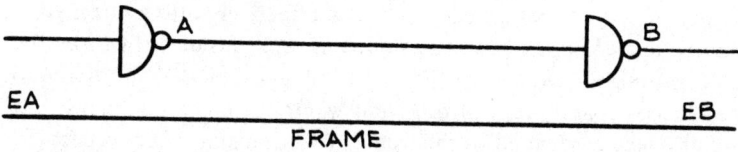

Figure 11.1

The logic gate at A (which can be regarded as a voltage source with very low output impedance) puts the correct logic signal v between A and E_A. After a short delay, this signal arrives between B and E_B. Let us now consider the whole machine, and the line AB within it (see figure 11.2).

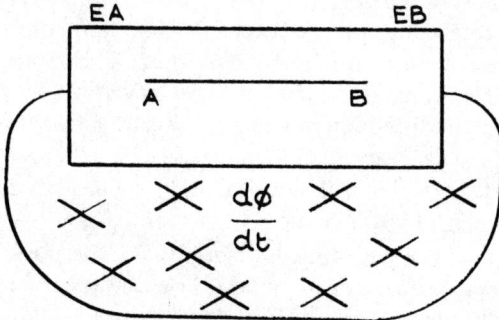

Figure 11.2

69

Let us suppose that the electrical interference causes a magnetic flux change $d\phi/dt$ to thread the loop formed by the frame of the machine and a ground loop, for example. By Lenz's law, this will result in (an eddy) current flowing around the loop in such a way as to oppose the change of magnetic flux. It will pass along the frame from E_A to E_B, and some will pass down the signal line AB. (An alternative reason for a surge of current would be if a human who was charged up with electricity touched the end of the machine, and the current was discharged from the human down the frame of the machine.) Because of the non-zero destination impedance of the signal line at B, and also due to the difference in physical size, etc., of lines AB and $E_A E_B$, the effect will not cancel out, and so the amplitude of the logic signal seen across BE_B will be affected. Clearly the more the reference line $E_A E_B$ could be made to look like the signal line AB, the better would be the rejection of such interference.

One obvious improvement would be to separate the logic 0 V from the machine frame as much as is possible (and legal; safety considerations indicate that all circuits should be tied down to ground). So if it is possible, we turn the machine frame effectively into a screened room, and keep all the logic and reference (0 V) inside as far from the frame as possible, except for the single link between 0 V and frame for safety reasons, preferably at the point where the incoming mains ground is also connected to the frame. The screening effect is best if the capacitance between the frame and the 0 V grid is kept to a minimum. This indicates that the amount of 0 V bussing should be kept down, a point which conflicts with the conventional wisdom on the subject of 'good grounding'. However, there is not really much chance of reducing the 0 V to frame capacitance very much, once the obvious precaution has been taken of avoiding whole surfaces of 0 V separated from surfaces of frame by only a few millimetres of insulation, and the total capacitance between 0 V and frame throughout the machine is then down to around 1000 pF or so. This represents about 200 Ω at 1 MHz, which is the kind of frequency at which we worry about externally induced interference. At this frequency therefore, if the 0 V and frame are separated as much as is practicable, the impedance between the two is still only 200 Ω in a typical machine which is one or two metres long and high.

By moving the 0 V and signal lines inside, as far as possible away from the frame, we have done as much as we can to ensure that externally induced interference flows down the frame only, leaving the logic unaffected. The next thing we can do is to consider whether we can nullify the effect of what little interference still does penetrate inside on to the 0 V and signal lines.

Let us assume that a large current, caused by external interference, flows down the 0 V bus. (We are thinking in terms of 1 MHz noise, and not very high frequency noise like 100 MHz. At 1 MHz, where the wavelength is 300 m, the picture shown in figure 11.3 is valid. That is, the wavelength is long compared with circuit dimensions, and the machine frame is part of a loop as in figure 11.3 rather than part of a transmission line, as it would be at 100 MHz and above.)

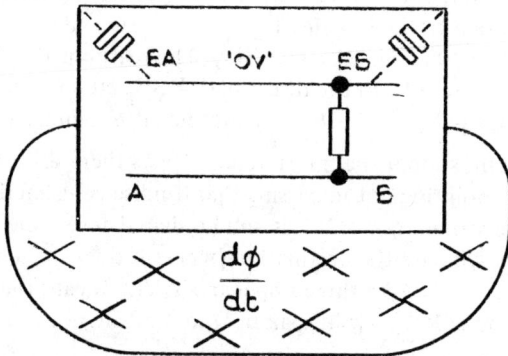

Figure 11.3

Due to resistance in the 0 V line between E_A and E_B, there is a voltage drop between them. The source impedance AE_A can be neglected, since the circuit driving the logic signal output is a voltage source. But the impedance BE_B — the destination impedance — is large compared with line resistances, so very little of the current caused by external interference flows down the signal line. This means that there is no voltage drop down the signal line to cancel out the effect of the iR voltage drop between E_A and E_B due to the ground current caused by externally induced current and the resistance of the ground line between E_A and E_B.

i = noise current
R = ground resistance
v = correct logic signal
v_{BE_B} = $v + iR$

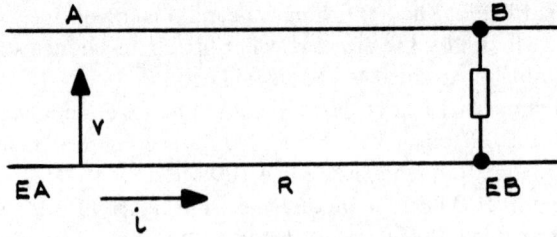

Figure 11.4

Figure 11.4 is a rough and ready description of what is really a complex situation, but it is sufficient to illustrate the point, and to show why noise into the 0 V line $E_A E_B$ is not cancelled out, but upsets the magnitude of the logic signal at BE_B. The reason is that the path $E_A ABE_B$ is different from the direct path $E_A E_B$, the former being a high impedance path (due to R_{BE_B}) and the latter being a low impedance path, so that the interference affects them differently.

It is worth pointing out in passing that if noise is magnetically induced as shown in figure 11.2, it will be shared across the source $E_A A$, the line AB, and the destination (where it does the damage) BE_B proportionate to the three impedances. This means that a *high* source impedance R_{AE_A} will 'soak up' more of the noise and keep it away from the destination. This contradicts the conventional view that a low output impedance for a logic gate is best for noise rejection. (However, capacitively induced noise will be best suppressed by a *low* output impedance.)

12

Grounding in a System Comprising more than One Module

The best approach to a system with more than one module is to extend the technique discussed previously for a one-module system.

The frames of the modules are connected together to form an extended 'screened room' (see figure 12.1). Inside the frames, the 0 V and the signal lines should be connected from one module to the other,

Figure 12.1

keeping the capacitance to frame to a minimum. Note that if the frames are connected together via a good shield around the interconnecting cable, the screening around the cable may be the best, but the capacitance between 0 V and the screen (that is, the frames) may be alarmingly high, perhaps 400 pF, or 400 Ω at 1 MHz. It follows that the best screen may not be best for the system, which demands not only good screening by the frames, but also minimum capacitance between 0 V and the screen. However, in practice this question turns out to be rather academic, because in any case it is hard to reduce the capacitance between 0 V and screen in the interconnecting cable significantly. However, we try to counterbalance the tendency for 0 V to be upset in the interconnecting cable by providing more noise rejection in signals going between modules than in signals within one module.

Safety in a System Comprising More Than One Module

The reader may have noticed an anomaly in the above discussion. Firstly, we say that the 0 V grid should be kept as far as possible from the frame, so that the frame will screen the logic from external electrical interference. Secondly, we concede that for reasons of safety, 0 V and frame must be tied together, so we rule that this should occur only once (although this has already marginally degraded the system, making it slightly more susceptible to interference; a completely floating electronics would be ideal from the point of view of noise rejection). Now we are considering a system with more than one module. Should 0 V and frame be connected together in only one module? If so, what happens to safety if two modules are disconnected, so that now one module's electronics has lost its earth strap?

The above is the first problem. The second problem arises if, as so often occurs, electrical power enters each module separately from outside. What do we do with the ground (third) wire coming into the second and third modules, and so on? Let us discuss these two problems separately.

Connecting 0 V to Frame in Systems Comprising More Than One Module

From the electrical point of view, the ideal solution is to have 0 V isolated from frame in every module except one, the main module. However, this is frightening from the safety point of view, since in most systems it is easy (and possibly necessary) to separate modules while still being able to supply power to them.

The worst possible situation arises if we try some sort of compromise, by tying 0 V to frame once, and only once, in every module. This is worse than strapping 0 V to frame everywhere, because there is a tendency for interference to be picked up by the frame (acting as an antenna) and then directed into specific areas of the logic. So if it is not possible to limit the number of connections between 0 V and frame in the whole system to one, they should be connected everywhere, so that the frame becomes the basis of the 0 V grid.

The reader might think that the latter course is the practical one in the circumstances, that is, to forget about trying to use the frame as a screen and revert to 'good grounding techniques' by tying 0 V to frame everywhere. However, he should be warned that the world is littered with systems where this has been done and where the performance is unsatisfactory — computers which keep failing for no apparent reason, or which fail when someone touches them. [A discussion of the mechanism of electrostatic discharge when a human touches a machine and its effect is discussed in T. S. McLeod and G. Johnson, 'Protection of Data Processing Equipment against Static Electricity Discharges', *Electronics and Power*, 24 (1978) 521–6.] If a machine is to work really satisfactorily, this problem has to be faced. A multiple module system, being larger than a single module, picks up even more electrical interference than a single module. The only viable solution is to have 0 V tied to frame only once in the whole system (unless the advice in the next paragraph is taken). The designer must also ensure that the current-carrying capability of the 0 V line between modules is sufficient to carry the current needed to blow a mains fuse in another module, should a fault occur in a subsidiary module so that there is a heavy surge of current into the 0 V grid.

A good way out of this dilemma is to tie 0 V to frame only through a choke (inductor). This is quite practicable. For instance, a 200 μH choke which will carry 20 A weighs only a kilogram or so, is only 100 mm cube in size, and costs around £5.

Specification of 0 V Choke   The choke must be able to carry a current in excess of the rating of the line fuses in the system, meaning something like 20 A. Further, it should not saturate at this current, because in the case of a human touching the 0 V line, and so discharging to ground through this choke, it should be able to absorb all of the energy stored in the electrostatic charge in the human before saturating, and so be effective in limiting the current and the rate of change of current through it, since the current causes troublesome iR drops across the 0 V grid, and the changing current causes troublesome electromagnetic fields. These requirements are met by a 200 μH choke which saturates at 20 A.

Ground Wires Entering More Than One Module

If line power is supplied individually to modules, and the ground (third) line is tied to the local frame every time, a second situation arises which causes designers some concern. This is because a complete loop is formed by the two ground lines and the frame (see figure 12.2).

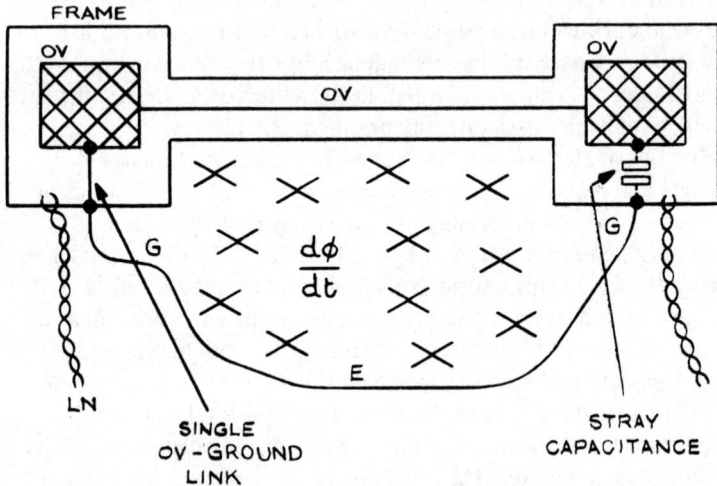

Figure 12.2

Should some stray magnetic flux $d\phi/dt$ happen to pass through this loop, then by Lenz's law, a current will flow around the loop which will tend to oppose this change of flux. Now we see the value of separating 0 V from frame, so that this current only flows in the frame and does not get near the logic, except via stray capacitance between 0 V and frame in module 2, which is kept as low as possible (say 1000 pF, or 200 Ω at 1 MHz). The path through the frame is very much preferred by this current over the high impedance path via the 0 V grid, so that 0 V throughout the system tends to stay at the same potential as the potential of the grounding point in module 1. (This last remark is a practical approximation to the truth rather than the complete picture, which is complex.) So we see that the existence of the ground loop is not very serious. (However, if 0 V had been tied to frame throughout the machine, it would be of much more concern, and the machine would be very susceptible to interference. This susceptibility could then be reduced by cutting all the ground lines into the machine except one, but here again we have a safety hazard.)

A good case can be made for the use of a choke to connect the ground line into the module frame. The discussion on values, cost and size is the same as in the case of the choke to connect 0 V to frame, which has already been dealt with.

13

Interference from the 50/60 Hz Line

Single phase power is supplied by a cable comprising three lines, live, neutral and ground (L, N and G). The live and neutral connect to the primary of one or more power transformers, and the ground line is connected to the frame of the machine for safety.

Interference from the line can be classified into three types.

(1) Balanced. The noise signal travels equally down the L and N lines, and the G line acts as a return path. It is called 'balanced' by electrical power engineers, and would be called 'common mode' noise by electronic engineers. Balanced noise was the line-borne interference of most concern in wireless telegraphy and other electronic activities which preceded digital electronics. It caused 'ground currents' which would upset high gain linear amplifiers.

(2) Unbalanced. The noise signal travels down the L line and back on the N line, leaving the G line unaffected. It is called 'unbalanced' by electrical power engineers, and would be called 'differential mode' noise by electronic engineers. Unbalanced noise was not of such concern in the past, because it tended to get lost (that is, suppressed) in the d.c. power supply. It can be shown that any complex signal travelling down the three lines L, N and G, can be resolved into a common mode component and a differential mode component.

(3) Line-borne radiated noise, both balanced and unbalanced. The noise enters the module frame via the L, N and G lines, where it then radiates directly into the logic.

Susceptibility of a Digital System to Line Noise

Differential mode noise on the L and N lines tends to be smoothed out at two points, both at the rough d.c. capacitors and also at the regulated d.c. capacitors (see figure 13.1). However, large value capacitors have significant 'series inductance' because of long connecting leads, so that some of the noise, if not suppressed before the transformer primary, will pass all the way through the power supply and cause transient variations of the logic power supplies which will tend to disrupt the correct operation of the logic. Screening the transformer will not help significantly, because at the frequencies we are most concerned about, in the region of 1 MHz, differential noise is fed through the transformer from primary to secondary by transformer effect, not via inter-winding capacitance.

Common mode noise gets through the transformer via inter-winding capacitance, so a screened transformer will very much help to suppress common mode noise at a very reasonable cost. (Typically, interwinding capacitance for an unscreened transformer is 100 pF. With screening, this falls to an effective 1 pF or so in the region of 1 MHz. At 1 MHz, 1 pF looks like an impedance of about 200 k Ω.) Otherwise, what noise does get through the transformer will tend to lift the positive voltage relative to 0 V and ground, and it will tend to lift the level of 0 V at some points on the 0 V grid compared with others. The use of a choke rather than a short in the safety link between 0 V and ground will of course help to render the logic immune to this noise, because all the logic will tend to move together. Another way of putting this is to say that common mode noise which *does* find its way through the regulated d.c. lines will then see three loads in series (see figure 13.2): the link between the G line and frame, the link between frame and 0 V, and the line carrying 0 V across the logic to the link. If the 0 V to frame to G line link has a high impedance (that is, is a choke), most of the noise will appear harmlessly across it. The d.c. resistance of the choke should be kept below the maximum allowed by the electrical safety standards. (BS 3861 section 6 indicates that the d.c. resistance of such a choke should be less than 0.1 Ω.)

However, if the 0 V to G link is low impedance, the impressed noise will devote itself to lifting the potential at one point of the 0 V grid compared with the potential at another. This, of course, will degrade the logic signals, and tend to cause system malfunction.

Line-borne radiated noise, which is radiated from the line wires

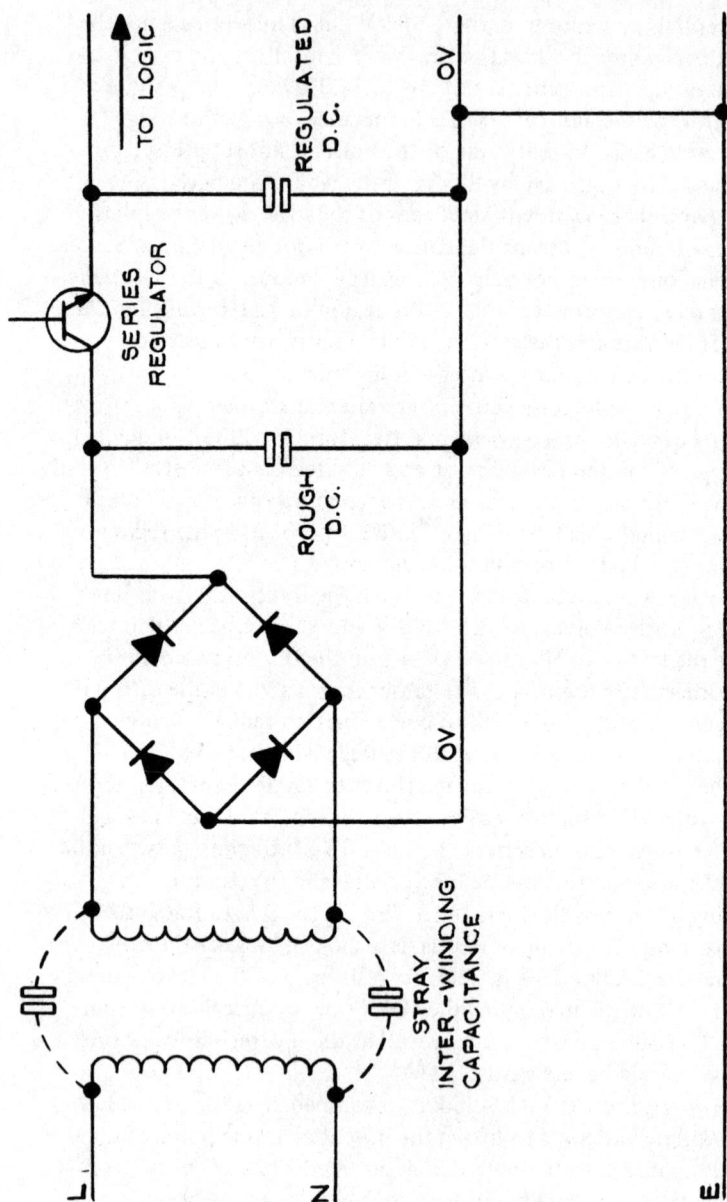

Figure 13.1

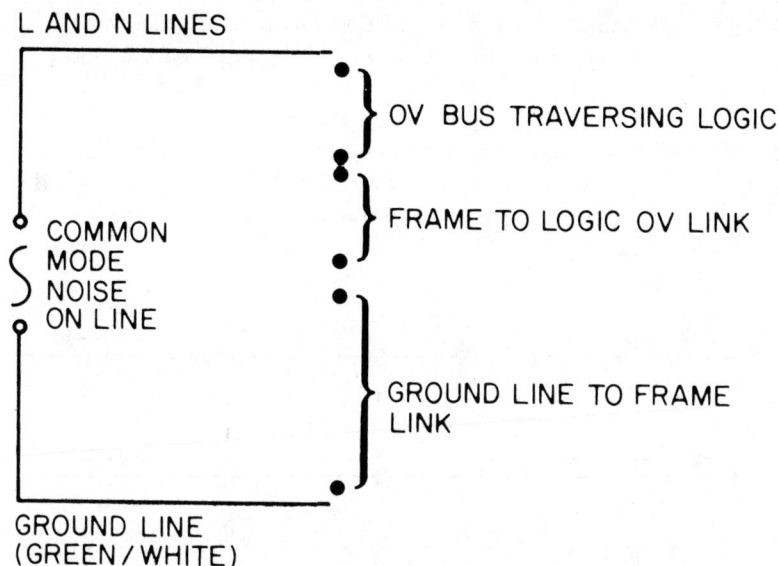

L AND N LINES

OV BUS TRAVERSING LOGIC

FRAME TO LOGIC OV LINK

COMMON
MODE
NOISE
ON LINE

GROUND LINE TO FRAME
LINK

GROUND LINE
(GREEN / WHITE)

Figure 13.2

inside the module, is greatly reduced by screening the L and N lines, the screens being grounded to the machine frame. Since screened cable is awkward to handle, it may be wise to try to avoid it. This can be done by having the line filter at the point where the power line enters the module, so that the supply lines within the module will be smoothed, and screening can be avoided. Another approach is to have supply lines in the module separated from the vulnerable logic by properly grounded bulkheads. Once past the line filter, the line cables need not be screened in normal circumstances. (That is, so long as power is never switched on and off to certain loads within the module, for instance compressors, while the logic is operating, switching transients in the mains should definitely be screened from the logic.)

Magnitude of Line-borne Interference

A reasonable noise amplitude to design against in a 240 V single phase supply is 2 kV over the range of 100 kHz to 10 MHz. (When thinking of the problem, imagine a pulse 2 kV high with a 1 μs leading edge and 10 μs width.) The noise may be common mode (down L and N and back on G) or differential mode (down L and back on N). It might be caused by the switching off of a 1 kW motor somewhere else on the same supply (see figure 13.3). When the switch is broken, there is a massive inductive kick from the load (motor windings) as it tries to maintain the current. This may pull current out of the G line via the capacitance between G and L. There will also be arcing (flashover) between the switch contacts, and so a complex voltage waveform results.

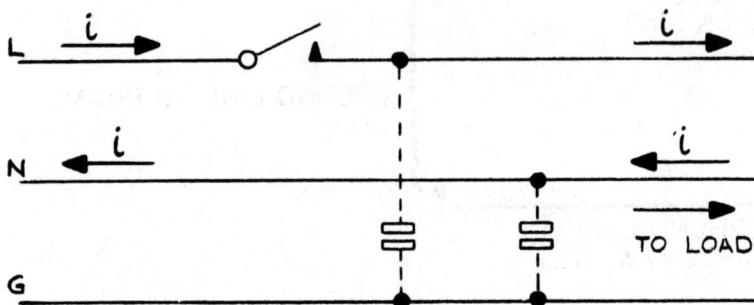

Figure 13.3

It is wise to assume large amplitude noise, definitely more than the nominal 240 V of the line, and also it is wise to assume both common mode and differential mode noise.

The source impedance of the noise is difficult to determine. It is safest to assume a very low source impedance, say 2 Ω. Both this conservative assumption, and also the assumption that the noise has an amplitude of 2 kV — which might surprise the reader, since this is something approaching an order of magnitude more than the nominal mains voltage — do not present the designer with a problem which is very different from the one he would have to face if he optimistically assumed a maximum noise equal to the nominal peak line value, about 350 V, and a 50 Ω source impedance, so he may as well play safe.

Useful confirmation of some key information in this chapter is found in *Transient Voltage Suppression Manual* (General Electric Semiconductor Products Department, 1976).

14

Filtering the Line

Line filters are made up of capacitors and inductors.

Capacitors   Those connected to the L line have 240 V across them, and so need to be rated for 240 V 60 Hz. They have to be able to dissipate the heat resulting from a rather surprisingly large current.
    By Ohm's law

$$V = IZ$$
$$I = \frac{V}{Z} = \frac{V}{1/6fC}$$
$$= 240 \times 360 \times C$$

For a reasonable value of capacitance, 1 μF, we find that the current turns out to be something approaching one-tenth of an amp. It is surprising to realise that a line filter can significantly alter the power factor of a load.
    The series inductance of such a capacitor, which incidentally is less than 300 mm cube in volume, can be as low as 10 nH, which is very satisfactory for our purpose. (At 1 MHz, 10 nH has an impedance of around 50 m Ω.)

Inductors (Chokes)   The important thing is to make sure that even at peak current the inductor is not saturated. If the power taken by the module is around 1 kW, so that the r.m.s. (measured) current is around 4 A, the peak current may be as high as 10 A. Noise must be suppressed during this peak current as much as at any other time.
    A choke which saturates at 20 A and has an inductance of 200 μH costs something like £10 and is about 100 mm cube in volume. Its

83

parallel capacitance can be as low as 10 pF, which is very satisfactory
for our purpose. (At 1 MHz, the impedance of 10 pF is around
20 k Ω.) The d.c. resistance of such a choke is around 0.1 Ω, so it will
be possible to meet safety requirements should the choke be put in
the ground line (in the United Kingdom, BS 3861 section 6a).

Types of Line Filter

A line filter is a low pass filter; that is, it allows through low
frequencies but blocks high frequencies. It contains series inductors
and parallel capacitors. The usual circuit is a double π (see figure 14.1).

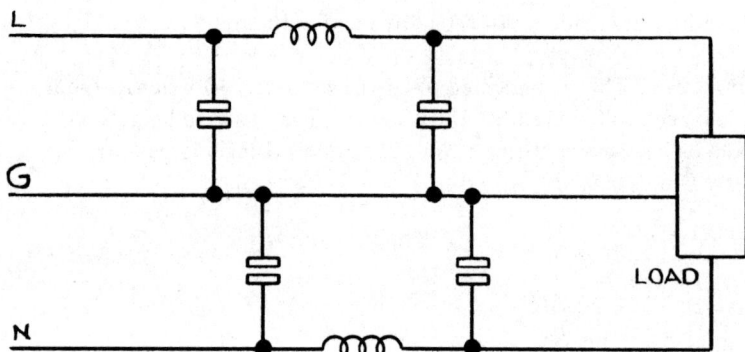

Figure 14.1

High frequency signals entering on either the L or N lines see a
high impedance inductance ahead and are shunted to ground through
a low impedance capacitor. Typically, at 1 MHz, with $C = 1 \ \mu F$ and
$L = 200 \ \mu H$

$$Z_C \approx 0.2 \ \Omega$$
$$Z_L \approx 1 \ k\Omega$$

If the source impedance of the noise is 1 k Ω or higher, this
attenuates the noise by a factor of

$$\frac{1 \ k\Omega}{0.2 \ \Omega} = 5000$$

or about 70 db.

If the source impedance of the noise is low, the first capacitor is ineffective, but the potential divider by the inductor and the second (downstream) capacitor still give around 70 db attenuation at 1 MHz, regardless of the impedance of the load.

Discussion of the Above Circuit

The filter is reassuring because it appears 'safe'. Any high frequency signal approaching from either direction sees a short to ground, which seems to shunt away the noise, and also sees a high impedance series inductor blocking its path ahead. If we were only worrying about noise coming down the line, it would be fine. However, it effectively clamps all the lines, both input and output, together at high frequency, so that the 'ground loop' pickup of externally radiated noise (see p. 76) can now also result from an L or an N loop, and so looks much more likely. Also, the possibility of electrostatic discharge into the module seems much more likely. From the point of view of radiated noise, the circuit in figure 14.2, which blocks the passage of high frequency signals down any lines, seems preferable. It makes the path down these lines an open circuit to high frequencies, and tends to isolate the system completely from its environment.

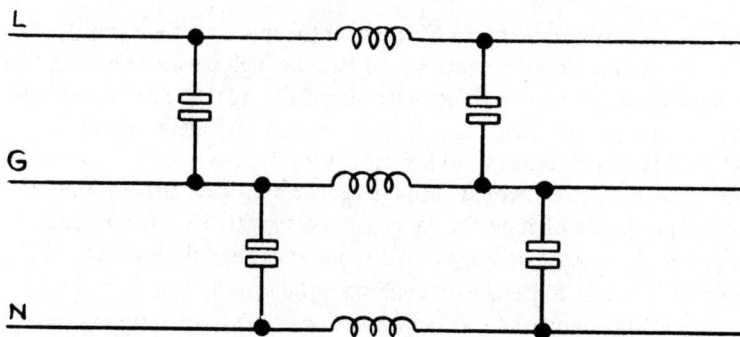

Figure 14.2

Ground Current

The above filter causes a disquieting amount of ground current. If the capacitors are 1 μF, the total ground current is about 150 mA. It seems much preferable to rearrange the lines as shown in figure 14.3.

The noise suppression is very little altered (being slightly improved

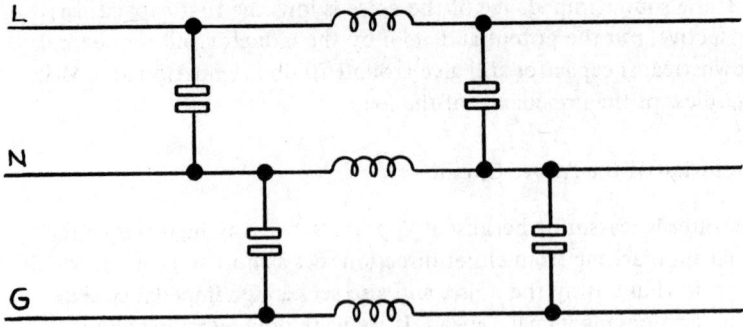

Figure 14.3

for differential mode noise and slightly degraded for common mode noise), and the ground current is reduced to a negligible amount, perhaps 2 mA. The circuit seems much safer too, because there are now no components linking L directly to G. This means that a single shorted capacitor presents no possibility of a safety hazard.

Line Filters on the Market

There are basically three types of filter on the market: the cheap (£10), the medium performance (£50) and the high performance (£200). The medium performance filters have about the right kind of specification – around 60 db insertion loss in the region of 1 MHz. A filter meeting this specification would cause 2 kV of noise to be reduced to a mere 2 V, which latter would easily get lost on its way through the power supply. The high performance filters, specified at 100 db insertion loss, reduce noise of 2 kV down to an unnecessarily low 20 mV.

The weak point in the above analysis is the assumption that the filters meet the manufacturers' specifications. The most serious shortcoming is when the windings of both chokes are on the same core. The idea is that the currents in the L and N lines, being equal and opposite, create zero total magnetic flux in the choke (see figure 14.4).

This means that for a heavy L and N current, the core will not saturate, and a single toroid costing £1 can be used in place of two separate chokes costing £10 each. Of course, instead of two chokes we now have merely a transformer, which will not stop differential mode noise at all. Unfortunately, even if the 50 Ω insertion loss is only

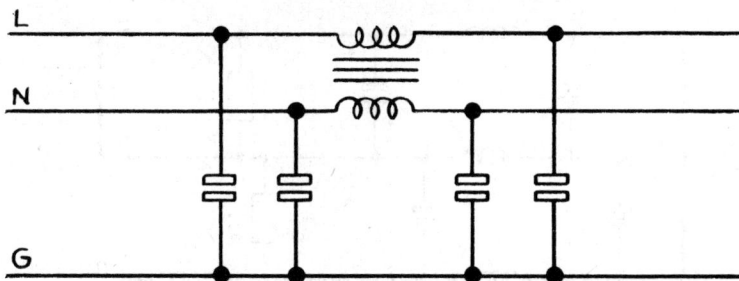

Figure 14.4

10 db for differential mode noise (down on L and back on N), manufacturers of such filters still feel entitled to claim 60 db insertion loss, leaving the unlucky customer to falsely assume that they claim this performance for differential mode noise as well as for common mode noise. The authors have never seen a manufacturer's specification where insertion losses for both differential (unbalanced) mode and common (balanced) mode noise were unambiguously defined. Further, 50 Ω insertion loss means the ratio of the amplitude of the output from the filter into a 50 Ω load divided by the input from a source with 50 Ω source impedance (see figure 14.5).

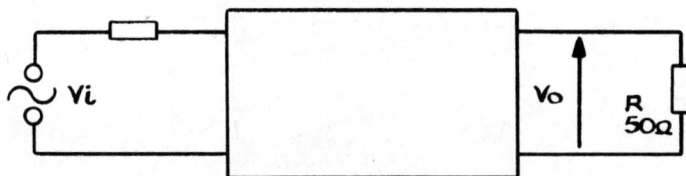

Figure 14.5

Sometimes, the manufacturer gives the 150 Ω insertion loss instead. However, usually he does not tell you what insertion loss is, but merely presents you with a single graph without explanation. Since the source impedance of the noise is difficult to determine, and the impedance of the load, certainly the instantaneous impedance (of a bridge rectifier, for instance), is virtually impossible to determine, the proper filter performance to specify is the 'minimum insertion loss', that is, the worst-case (least) attenuation when the source impedance

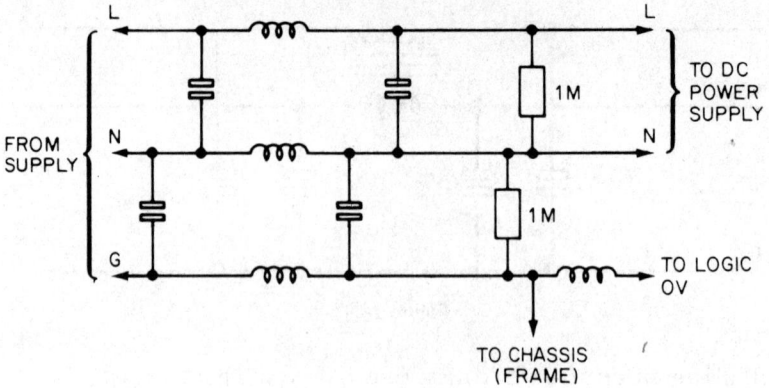

Figure 14.6

and load impedance are independently varied from zero ohms to infinity.

Figure 14.6 shows the line filter designed to meet all the considerations discussed above. (For information on the availability of these digital system line filters, write to Icthus Instruments Ltd, Princesway, Team Valley Estate, Gateshead, Tyne and Wear.)

15

Oscilloscopes for Digital Measurements

The oscilloscope is still the most commonly used signal analysis instrument available to the digital engineer. It is therefore relevant to consider its use carefully so that the maximum amount of information can be obtained from the display. The applications of the oscilloscope for digital signal analysis fall into two main categories

(1) checking the timing relationships of digital signals to verify the correct operation of the logic system,
(2) measurement of noise levels, checking clock edges, risetime measurements, etc.

For application (1) an oscilloscope with bandwidth in the range 50 to 100 MHz (risetime 8 to 4 ns) will usually suffice, whereas applications in the second group usually call for a higher bandwidth and may require the use of a sampling oscilloscope (typical bandwidth 1 GHz; 350 ps risetime).

The engineer can usually assume that the oscilloscope itself gives a reasonably accurate picture (subject to risetime limitations) of the signal presented to its input. In the authors' experience the chief problem area is associated with the means for conveying the signal from the circuit under test to the input of the oscilloscope. This brings us to the subject of probes.

High Impedance Probes

By far the most common type of oscilloscope probe is the high impedance 10:1 probe. This is designed to feed a standard 1M oscilloscope input; the circuit is shown in figure 15.1.

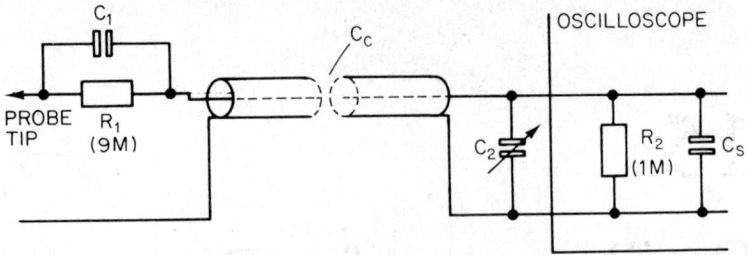

Figure 15.1

The basic principle of operation can be understood by noting that resistors R_1 and R_2 provide a 10:1 attenuator for signals at low frequencies where the effects of the various capacitances are negligible. At high frequencies C_1 together with C_C (cable capacitance), C_2 (trimmer capacitor) and C_S (oscilloscope input capacitance) form a capacitive divider. For correct operation of the probe it is necessary that this capacitive divider should give the same attenuation as the resistive divider. Since capacitive reactance at frequency ω is given by $Z_C = 1/\omega C$ we have

$$\frac{R_1}{R_2} = \frac{Z_{C_1}}{Z_{(C_C + C_2 + C_S)}}$$

that is

$$\frac{R_1}{R_2} = \frac{C_C + C_2 + C_S}{C_1}$$

In practice C_2 is adjusted until this condition obtains. The normal method is to observe a square wave signal and adjust C_2 for zero undershoot and overshoot.

It will be noted that the theory of operation rests on the assumption that the cable capacitance can be treated as a lumped component. Once the transition time of a step signal along the connecting cable becomes longer than the risetime of the step this assumption breaks down and distortions of the pulse edge will be fed to the scope.

Various 'dodges' such as the replacement of the cable core by resistive wire have been used although, obviously, they cannot be entirely successful since the whole theory of operation of the device is based on a fiction.

In addition to the above mentioned drawback the 10:1 high impedance probe has an even more serious deficiency: it is extremely prone to pickup by virtue(?) of its high impedance. The user can easily convince himself of this by means of the following simple test.

Use the ground lead of the probe to ground the pulse tip so that the whole probe assembly forms a loop of 100 to 150 mm diameter, then place the 'loop' near to an operating digital board. Spikes will be observed on the oscilloscope. It transpires therefore that any signal which is now injected between the probe tip and ground lead will be in series with this direct pickup, thus the display on the screen will be a composite of the signal under study plus spurious noise. It should be clear by now that this makes it impossible to tell whether some of the features of the displayed signal are in fact genuine, and that the usefulness of the high impedance probe to the logic engineer should be seriously questioned.

The high impedance probe originated in the days of valve technology when the high impedances of the circuitry used (up to megohms) called for very high impedance measuring devices. In high speed logic impedance levels are drastically lower, of necessity, and hence the high impedance probe has ceased to be required. It should be allowed to slumber in peace, its *raison d'être* having been removed.

Low Impedance Probes

The correct probe to be used for high speed digital signals is the low impedance probe which will be described in this section.

The low impedance probe makes use of the following simple principle. Any resistor, R, in a circuit may be replaced by a transmission line of any length but of impedance Z_0 ($= R$) terminated in resistance R. This will cause no perturbation whatever to the circuit and will allow the signal which would have appeared across R to be observed remotely. The only penalty to be paid is the time delay necessary for the signal to propagate along the length of transmission line. In practice this is no disadvantage; it simply requires that when multiple points are being observed equal lengths of transmission line should be used. Figure 15.2 gives an example of what is meant by the above.

The emitter load resistor (figure 15.2a) has been replaced by a 50 Ω co-axial cable terminated in 50 Ω (figure 15.2b) and the signal across the 50 Ω resistor is displayed at the oscilloscope.

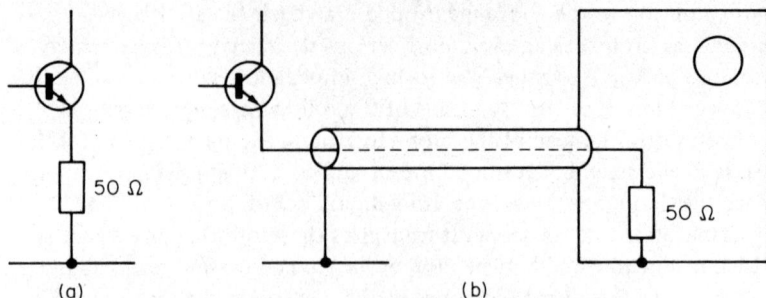

Figure 15.2

This technique is easily extended to values of resistor higher than 50 Ω as shown in figure 15.3, where the correct 2 kΩ resistor is effectively replaced by 1950 Ω in series with 50 Ω. The signal displayed by the oscilloscope is 1/40 of that which would appear across the 2 kΩ resistor.

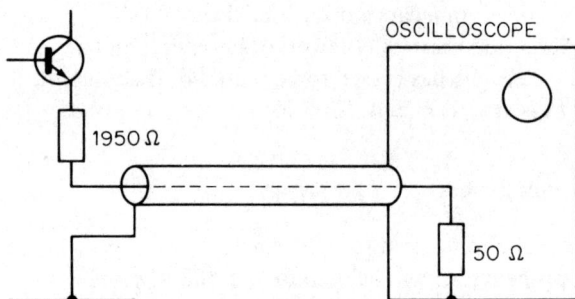

Figure 15.3

It is possible to use this technique to provide built-in probe points leading to co-axial sockets which normally carry 50 Ω terminators. In order to observe any of these signals all that is necessary is to remove the terminator and couple to the 50 Ω input of the oscilloscope with 50 Ω co-axial cable.

So far we have not described a probe as such but rather oscilloscope connection techniques. It is simple, however, to build a probe based on the principles just described.

Figure 15.4 shows the constitution of such a probe and figure 15.5 is a schematic diagram.

In order to minimise loading of the circuit under observation R_1 should be made as high as possible considering the noise level of the

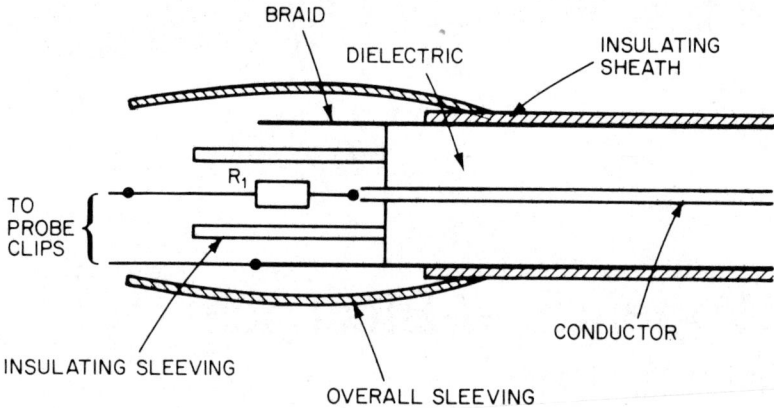

Figure 15.4

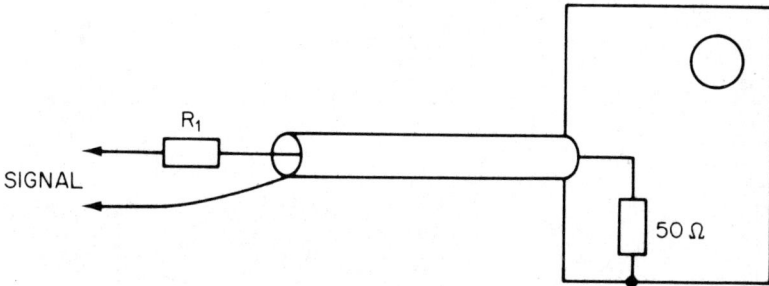

Figure 15.5

oscilloscope in use. The wire ends of the probe must be kept as short as possible and ideally soldered to the circuit under test. In practice this is inconvenient and so miniature prodclips (ezi-hook) can be fitted. These are only about 20 mm long and the wire leads should not exceed about 25 mm long.

If the oscilloscope used does not have a 50 Ω input, a 50 Ω terminator should be used between the co-axial cable and the oscilloscope.

Grounding of Probes

Whatever type of probe is in use it is essential that its ground lead be connected to circuit ground as close as possible to the circuit point under observation. Where more than one probe is in use this rule should be observed for each probe.

Appendix—Electrical Formulae

	CAPACITANCE	INDUCTANCE	CHARACTERISTIC IMPEDANCE
Long Straight Wire		External $\quad L = \alpha$ H/m Internal $\quad L = \frac{\mu_0}{8\pi}$ H/m	
Concentric Conductors	$C = \frac{2\pi \, \epsilon_r \epsilon_0}{\ln \frac{R}{r}}$ F/m	High Frequency $L = \frac{\mu}{2\pi} \ln \frac{R}{r}$ H/m Low Frequency $L = \frac{\mu}{4\pi} \left(\frac{1}{2} + 2\ln \frac{R}{r}\right)$ H/m	$Z_0 = \frac{1}{2\pi} \sqrt{\left(\frac{\mu}{\epsilon}\right)} \ln \frac{R}{r}$ $= \sqrt{\frac{L}{C}}$
Parallel Wires $\quad a \gg r$	$C = \frac{\pi \epsilon_r \epsilon_0}{\ln \frac{a}{r}}$ F/m	Self-inductance High Frequency $\quad 2L = \frac{\mu}{\pi} \ln \frac{a}{r}$ H/m $= 0.4 \ln \frac{a}{r} \, \mu$H/m Low Frequency $\quad 2L = \frac{\mu}{4\pi} \left(1 + 4 \ln \frac{a}{r}\right)$	$Z_0 = \frac{1}{\pi} \sqrt{\frac{\mu}{\epsilon}} \ln \frac{a}{r}$ $= \sqrt{\frac{L}{C}}$
Wire Above Ground Plane $\quad a \gg r$	Twice Value for Parallel Wires	Half Value for Parallel Wires	Half Value for Parallel Wires
Parallel Plates $\quad a \gg b$	$C = \frac{\epsilon a}{b}$ F/m	Self-inductance $2L = \mu \frac{b}{a}$ H/m	$Z_0 = \sqrt{\left(\frac{\mu}{\epsilon}\right)} \times \frac{b}{a} = 120\pi \sqrt{\left(\frac{\mu_r}{\epsilon_r}\right)} \times \frac{b}{a}$
Pair of Wires, Ground Return	Mutual Capacitance (= 1/Coefficient of Induction) $D = \frac{\epsilon \times 2\pi}{\ln \frac{\sqrt{(4h^2+d^2)}}{d^2}}$	Mutual Inductance $M = \frac{\mu}{2\pi} \ln \frac{\sqrt{(4h^2+d^2)}}{d^2}$ $= \frac{\mu}{4\pi} \ln \frac{4h^2+d^2}{d^2}$	$\mu_0 = 4\pi \times 10^{-7}$ $\epsilon_0 = \frac{1}{36\pi} \times 10^{-9} = 8.85$ $c = \frac{1}{\sqrt{(LC)}} = \frac{1}{\sqrt{(\mu \epsilon)}}$
	For Close Pairs	$M = \frac{\mu}{2\pi} \ln \frac{ad}{bc}$ $M = \frac{\mu}{2\pi} \ln \frac{bc}{r^2}$	$\frac{1}{\sqrt{(\mu_0 \epsilon_0)}} = 3 \times 10^8$ m/s
Two Signal Wires, Common Return		$M = \frac{\mu}{2\pi} \ln \frac{b^2}{ar}$